GOLF IN THE LOWCOUNTRY

An Extraordinary Journey
Through Hilton Head Island & Savannah

Joel Zuckerman

with a foreword by Rees Jones

To Rick—
All the best!

Joel Zuckerman

MORE PRAISE FOR
GOLF IN THE LOWCOUNTRY

"Joel Zuckerman has been a contributing writer for my web site Shark.com for years, and his features on courses throughout the world have always been entertaining. His latest work, *Golf in the Lowcountry*, takes a close look at one of the country's most underrated golf regions, and with his home in nearby Savannah, he is uniquely qualified to expound on the great courses of the area. I hope you enjoy this book as much as we all enjoy playing there."
  – Greg Norman, Eighteen Time PGA Tour Winner, Two-Time British Open Champion

"Joel Zuckerman has been interviewing me for a long time. He understands the game, is a fine writer and knows all there is to know about golf in the Lowcountry. *Golf in the Lowcountry* is a great read."
  – Billy Andrade, Four Time PGA Tour Winner

"Zuckerman knows his golf, knows the Lowcountry and can play a bit besides. He laid four natural birdies on me the last time we played a round together. Call him 'the Buddha of the Bermuda.' "
  – Loren Roberts, "The Boss of the Moss," Eight-Time PGA Tour Winner

"I've lived on Hilton Head for more than 30 years, and am quite familiar with all the great tennis and golf here in the Lowcountry. Joel Zuckerman also knows this area well, and all the wonderful courses that are here. He has a good game and a terrific style in describing these treasures, as well as some of the personalities that live and play here in *Golf in the Lowcountry*."
  – Stan Smith, former #1 Tennis Player, U.S. Open and Wimbledon Champion and 30-year Hilton Head Island resident

"This is the perfect book for the low-attention span golf nut, headed for the Lowcountry. Read it from cover to cover, or start anywhere you want, it's filled with all you need to know about Lowcountry golf courses, and much, much more. Take it off the coffee table, and stick it in your hand luggage!"
  – David Feherty, CBS Television

"Joel Zuckerman is a fine golfer and a bright fellow, but why a nice Massachusetts boy like him would leave God's country in New England to head to the steamy South to live was one question he never answered for me satisfactorily. Until now. His book is loaded with convincing answers, from the immediate appeal of the Lowcountry golf courses, to the more timeless attraction of the region contributed by its people and exotic landscapes. Zuckerman has put down roots in fertile soil for good golf and good golf writing."
  – Tom Bedell, *American Airlines' Celebrated Living Magazine*

"Few golf writers know Savannah and the Hilton Head area as well as Joel Zuckerman and it shines through in *Golf in the Lowcountry*. Witty, lyrical and informative, he captures, in a refreshingly entertaining style, all the qualities that make the region a golf haven. From vibrant course descriptions of revolutionary Harbour Town and unforgettable Haig Point, to a thought provoking profile of two-time U.S. Open champion Payne Stewart, this book has been long overdue."
  – Russ Christ, Editor, *Arizona, The State of Golf*

"Joel Zuckerman knows his sport and even more, he knows his turf. His book will be a great help to golfers in the Lowcountry."
  – Kevin Cook, Managing Editor, *Travel & Leisure Golf*

"Too many golfers forget that it's just a game. Joel Zuckerman offers insight into golf with a sense of humor that's much needed. His keen observations of Lowcountry golfers, golf courses and the game itself are all included in this fine book."
  – Ryan Rees, Editor, *Georgia Golf News*

"When it comes to golf in the Lowcountry, Zuckerman knows it like he knows the back of his 9-iron, which he uses repeatedly taking divots in the wonderful courses of the region. *Golf in the Lowcountry* will inspire you to want to do the same."
  – **Jeff Thoreson, Editor, *Washington Golf Monthly***

"He admittedly hadn't written so much as a grocery list in more than 15 years, but I had a good feeling about Zuckerman when I hired him to be our golf writer. My instincts were correct, and I'm not surprised that his byline quickly appeared in magazines across the nation. His irreverent perspectives on the courses and characters around Hilton Head have struck a chord with our readership, and they'll do the same for you."
  – **Kyle Poplin, Editor, *Carolina Morning News***

"After spending one too many late evenings listening to him holding court, I finally told myself, 'There's got to be an easier way to find out what this guy knows about the Lowcountry.' This is it. I find reading Zuckerman so much more enjoyable than listening to him, I'm crossing my fingers this isn't scheduled for audio-cassette."
  – **Reid Champagne, *Chicken Soup for the Golfer's Soul***

"Joel Zuckerman is a bright, entertaining, funny...and deeply disturbed man. In his new book, *Golf in the Lowcountry*, he reveals fully the first three attributes and just a hint of the latter. His approach to life and writing, combined with the undeniable attractions of golf, has resulted in a book no golfer should be without."
  – **John Gordon, *Affluent Golfer Magazine***

"Joel Zuckerman has an engaging writing style that will have you nodding your head in agreement and laughing out loud at his observations about this great game we play. This Yankee transplant, I believe, has been sucked into the lure of southern golf. After you've read *Golf in the Lowcountry*, I think you'll agree."
  – **Judy Putnam, Editor, *Georgia Fairways***

"As a fellow Lowcountry golf aficionado, I am pleased our little neck of the woods will receive recognition for the excellence not only of its golf courses, but of its lifestyle in general. Despite his immutable status as a transplanted Yankee, Joel Zuckerman is sure to make his island digs feel like home."
  – **Brad King, Senior Editor, *LINKS Magazine***

"Joel Zuckerman has consistently provided us with thoughtful and informative stories on Lowcountry golf through the years, and why shouldn't he? He has both the knowledge and enthusiasm about the area, and the ability to communicate it to our readers."
  – **Linda Wittish, Editor, *Savannah Magazine***

"Zuckerman is the Elvis of Hilton Head/Savannah golf writers, all others are mere impersonators. He is a hound dog for the best courses. No need to get all shook up figuring out the best places to play. Learn from the King!"
  – **Larry Miller, Editor, *Coastal Golf Magazine***

"A thorough and entertaining look at one of the most beautiful and golf-rich coastal regions this side of Scotland. Enticing, fun and authoritative."
  – **Brian McCallen, Senior Editor, *GOLF Magazine***

"Zuckerman truly knows his subject matter. When he's talking Lowcountry you can sniff the golf. When he's talking golf, you can sniff the Lowcountry."
  – **Brian Hewitt, The Golf Channel**

*Golf in the Lowcountry, An Extraordinary Journey
through Hilton Head Island and Savannah*, Copyright 2003
by Joel Zuckerman

All rights reserved. No part of this book may be used,
reproduced in any manner whatsoever without written
permission except in the case of brief quotations embodied
in critical articles and reviews. For information contact Saron
Press, Ltd., P.O. Box 4990, Hilton Head Island, SC 29938.
info@saronpress.com

Produced by Saron Press, Ltd.
Edited by Paul deVere
Designed by Michael Reinsch Design
Cover art and illustrations by Brenda Turner, Copyright 2003
Printed by Regal Printing, Hong Kong, China

First Edition

Library of Congress Control Number: 2002115572
ISBN 0-9650791-6-3

Special note. Personality profiles and essays are used through
the courtesy of *Carolina Morning News.*

For more of Joel Zuckerman's writing visit www.vagabondgolfer.com.

TABLE OF CONTENTS

INTRODUCTION

It's not hard to understand the appeal of the Lowcountry, roughly defined within these pages as the coastline and adjacent interior lands between Hilton Head Island, South Carolina, and Savannah, Georgia. Anywhere you can consistently drive a convertible with the top down you can rest assured that the quality of life is going to be way up.

But it's more than the temperate climate, lush vegetation and mellow pace of life that are hallmarks of this special environment. Granted, the natural landscape is extraordinary. Golden tidal marshes, gnarled, centuries-old live oaks draped with Spanish moss, forests of hardwood and wide stretches of pristine beach captivate both travelers and long-time residents alike. No one disputes the beauty of the views or the friendliness of the people, but in a word, recreation is the key attraction of the area. The abounding waterways; rivers, creeks, inlets and lakes in addition to the ocean, are a boater's paradise. Fishing, crabbing, kayaking, snorkeling, windsurfing, parasailing and swimming are popular pursuits. Back on land, beaching, biking, roller-blading, and tennis are practiced all year long, with no shortage of joggers, kite flyers and horseback riders in evidence as well.

These pastimes are all perfectly fine, but to me, the essence of the Lowcountry is golf. There are plenty of folks who love living here or visiting and never take club in hand, but other than my wife and kids, I generally choose to ignore them. To my mind it's the dozens of spectacular courses that make this region so inviting. From the subtleties and stunning vistas of Harbour Town Golf Links to the rustic forests of Daufuskie Island's Melrose Course, just a short boat ride away. It's the dramatic bunkering of Pete Dye's effort at Colleton River, the heroic par 3s at Arnold Palmer's Crescent Pointe, the unforgettable view of the Savannah River Bridge from the fairways of The Club at Savannah Harbor, and the mind-boggling array of golf holes at The Landings Club that make the region so unforgettable from a golfer's perspective.

The vignettes contained herein discuss these courses and many others, along with profiles of well known golfers here in the Lowcountry, and better known golfers recognized in any country. Perhaps you'll recognize yourself or someone you golf with regularly in the essays found within, humorous situations and subjects that are as endemic to the game as tees, Titleists and Top-Flites.

If whacking a white ball through a green meadow strikes you as a good time, read on. Like me, there are thousands of you who ultimately feel that golf is indeed the lure of the Lowcountry.

DEDICATION

Dedicated to 6 gals and the memory of a man.
To my mother, wife and daughters, who have never played the game, and to my sisters,
who look as though they could make the same claim. And to my late father Karl,
a remarkable man whose passion for golf only inspired boyhood indifference,
but also a latent fascination years in the making.

ACKNOWLEDGEMENTS

This book is the end result of an impulse phone call I made in the autumn of 1997. We had recently landed in the Lowcountry from New England, looking for more fun and more sun than we were accustomed to up north. Idly perusing the classified ads, I came across a listing that announced: "Golf writer wanted for Hilton Head newspaper."

Intrigued, I arranged for an interview. I thought I knew how to golf and how to write, although it took awhile to realize how inadequate my preparation was on both counts.

My freelance position at the *Carolina Morning News* was well named. I was free to create essays, contests, profiles, course reviews and the like, with little interference. Because I didn't know any better, about six months into the position I cold-called *Sports Illustrated* with a story idea. I managed to get an editor on the telephone, and my manic spiel convinced him to assign me a 250 word brief. Either that or he just wanted to get off the telephone. In any case, I had a toenail in the door at Time/Life.

Eventually I moved on to bigger pieces, longer features and more in-depth profiles at magazines and websites like *Golfweek, Golf International*, travelgolf.com and pgatour.com among many others. Editors such as Cameron Morfit, Dan Gleason, Brad Klein, Brian Hewitt, Joe Flynn and Kevin Cook have been helpful to me. So has Paul deVere, my publisher on Hilton Head, who was instrumental in turning this rough idea into reality.

The area golf pros have been generous in providing access to their golf courses. I'm grateful to Scott Stillwell, Jim Mancill, Franklin Newell and Mike Harmon among others for granting repeated access to their facilities without expecting a five or six figure initiation check in return.

The vibrant cover art and thoughtful illustrations were created by a wonderful artist named Brenda Turner, to whom I am indebted.

I must also thank my lovely wife Elaine, who remains bemused if not enthused as I traipse about, ostensibly in the name of "business," on more golf excursions in a year than most avid players will take in five.

I should pay brief homage to my personal golf posse as well, consisting of Sandy Andy, Reload Richie, Easy Ed, Captain Rivi and the munificent Meeglemoon, among others. We've enjoyed countless stick and ball safaris in the Lowcountry and points well beyond, and the boys never fail to amuse, amaze, annoy and inspire -- often on the same hole.

Lastly I must thank Kyle Poplin, the editor of the *Carolina Morning News*, who hired me at the beginning. We've butted heads so often over the years my hairline has a permanent indentation. But I'm perpetually grateful he had the foresight to give me a shot, and I always tell him how lucky I was to get what I assumed to be a coveted position. Self-effacing to the end, his response is always the same. I assume he's kidding, but his face is inscrutable when he says, "It's really no big deal, Joel. It had nothing to do with luck, or skill either, for that matter. The truth of it is you were the only one who applied for the job."

FOREWORD – *By Rees Jones*

It's been almost thirty years since I struck out on my own as a course designer. I've been fortunate to have been involved in well over a hundred projects during that time, either creating original works or renovating classic courses, in many instances preparing them for Major championships.

I've worked in 31 different states and several other countries, but I find myself returning time and time again to South Carolina's Lowcountry and the Georgia coast. There is something about the area that has always intrigued me, ever since I made my first visit to Hilton Head Island in 1969. Back then, there was little in the way of development, just a few resort hotels and a smattering of golf courses. But I could see immediately the potential that existed for creating world class golf, a sentiment that's obviously been shared by many of my colleagues in the ensuing decades.

The Lowcountry is an intriguing prospect for a golf course architect. On the one hand, there is little elevation change available. We must create and integrate design features that didn't exist to begin with, and do so in a manner that will appear natural and uncontrived. But the region also abounds with magnificent physical attributes – dramatic ocean views, vast stretches of salt marsh, forests of slash pine, centuries-old live oak and ever present palmetto trees.

The Lowcountry and coastal Georgia have been the canvas for many notable architects who have created dozens of fine courses, many of which are described vividly in this book. Joel Zuckerman travels, writes about and rates golf courses extensively. I know this for a fact, as we've run into each other on many occasions, both in the Lowcountry and beyond. He has a firm grasp of the subtleties and nuances that distinguish the better designs. He is able to convey these differences in a style that is both entertaining and informative ... no easy task. In his travels, Joel has also become acquainted with many of the more notable players in the area, and the profiles of both local and national golf figures he's encountered make for interesting reading as well.

To top it off, Joel has added a third element that makes *Golf in the Lowcountry* as unique as the courses he has written about. Throughout the book, you will find his observations about the game itself. His essays aren't necessarily endemic to the region, but instead speak to the universal golf experience with comedic insight, and are among the biggest strengths of this book. But that's only my opinion, and ultimately up for you to judge for yourself. I can say this volume is well titled, though. I've done some of my finest work in this region, and I will say that it has been "an extraordinary journey."

SECTION I:
HILTON HEAD ISLAND

COURSE DISCOURSE:
HARBOUR TOWN GOLF LINKS

Number 18. Courtesy of Sea Pines Resort

There are Parisian tourists, either time-challenged or acrophobic, who miss the Eiffel Tower. Certain visitors come to New York and don't take in a Broadway show, and likely a select number of travelers to Egypt who miss the Pyramids at Giza. While it's still possible to enjoy a full volume vacation, it's arguable that these vacationers are missing the highlight, the essence of the area by avoiding or eschewing these destinations.

By the same token, it's entirely possible to have a great golf vacation in the Lowcountry and never set foot on the Harbour Town Golf Links. There are dozens of daily fee options, both on and off the island, that provide an excellent challenge and notable scenery, and are available at a fraction of the Harbour Town greens fee, which can be as high as $250. But Harbour Town is the pinnacle of golf on Hilton Head, well within the top 20 modern courses in the nation according to Golfweek Magazine, and one of the most inventive routings of the modern era. This Pete Dye-Jack Nicklaus collaboration is truly a course for the ages.

"It's one of the most innovative and revolutionary designs in the history of architecture," states Brad Klein, Golfweek's architectural editor. "It's certainly one of the ten most important courses in terms of design, because Pete Dye built all sorts of great contour, shape, form and strategy into a dead level site that was really quite boring to begin with. Instead of moving massive quantities of dirt, he massaged the earth in a subtle way, turning the holes, and positioning them so the live oaks draped the entrances to the green. It created a tremendous sense of corridors, and you have to keep working the ball from left to right and right to left. It's a really ingenious design. Here's this dead flat golf course with tiny little nothing greens, and it drives you nuts. You've got to love it."

There's much to recommend, indeed much to love about Harbour Town beyond the tiny greens, dynamic bunkering and spectacular finishing holes that are often the first attributes mentioned. The serpentine course routing offers an excellent variety of driving holes. The first, third, eighth, ninth, eleventh and twelfth, all par 4s, are tight to the point of claustrophobic when viewed from the tee. By contrast, the second, fifth, tenth, sixteenth and eighteenth holes all offer the appearance of generous landing areas. But even here, in some cases an improperly positioned tee shot can be in the short grass yet still provide no clear route to the green.

It's virtually impossible to recollect any other world class course that has the proliferation of houses and condos that are seen here at the southern end of the Sea Pines Resort. It's a testament to Dye's acumen then that a round here isn't at all like a typical ride through a neighborhood subdivision, which is so often the case in the region. The strategy required on each shot and the omnipresence of the fabulous hardwoods defining and influencing the line of play draw the attention. The housing and road crossings fade into the background, as players turn their concentration to the not so easy task of negotiating the golf course.

It's one of the most revolutionary designs in the history of architecture.

There's much consideration given to the par 3 holes here, and rightfully so. Three of the four are routed in different directions, and while they average less than 175 yards each from the blue markers, they are formidable nonetheless. Number four features water and a green with a bulkhead. The seventh has water and a Kalahari-like expanse of sand. The fourteenth is again over water, and has the nastiest little pot bunker behind the green, invisible from the tee, that is too harsh even for Pine Valley. Lastly there's the spectacular seventeenth, where wind, water, wetlands and a stunning view of the Calibogue Sound can conspire against even the most focused player.

One of the raps on Harbour Town was that it played either too short or too long for most golfers. Before the course renovation in 2000, a player's options were either the back tees at nearly 7,000 yards, (146 slope) or the white markers just over 6,000 (130 slope). Many guests didn't want to feel cheated after doling out such a hefty greens fee, and chose to play the championship markers. This resulted in slow play, frustrated patrons and plenty of lost balls. The single smartest move made by management was instituting blue markers at 6,600 yards (139 slope) when the course was reopened in 2001. This is the perfect distance for many players and has resulted in a golf course that provides an appropriate tee box for all levels of play.

Much of Hilton Head's great golf is found at the area's myriad private clubs, and is inaccessible to the public at large. It's nice to know that Harbour Town Golf Links, as fine as any other course in the Lowcountry, is available to anyone who chooses to meet its relentless challenge. It goes without saying that it's the finest resort property in the region. What's also true is that it might well be the finest course in the region, period.

Personality:
The "Big 3" of the Heritage

The weather was absolute perfection as I made my way into Sea Pines Resort, and served as a reminder to why we've chosen to make our home in the Lowcountry. With bright blue skies, temperatures barely cresting 80, and a gentle breeze wafting, the timing could not have been better to visit Harbour Town, Hilton Head's preeminent golf course. I was fortunate to have Harbour Town's "Big Three" join me on the links; Head Professional John Farrell, Superintendent Gary Snyder, and The Heritage Tournament Director Steve Wilmot. The only person missing was Director of Sports Cary Corbitt, but his presence would've ensured my absence, so I have to say things worked out just fine.

This harmonic convergence of Snyder, Farrell and Wilmot together on the greensward was a rare occurrence. How rare, you wonder? It was the first time _ever_ that they teed it up in unison. The gregarious Farrell offered a facetious explanation, saying "I try and avoid these guys whenever possible." The truth of the matter is that these men, all at the top of their chosen professions, long

ago learned the "dirty little secret" that comes as such a shock to every young assistant pro, range rover or bag boy in their first job. When you work in the golf business, there usually isn't much time for playing golf.

This harmonic convergence of Snyder, Farrell and Wilmot together on the greensward was a rare occurrence.

Steve Wilmot is still in his early 40s, but has already worked more than 15 tournaments on Hilton Head, five as the head man. His unrelenting work ethic serves to bother his conscience just slightly on this early autumn afternoon, as he makes occasional reference to office activity he was missing. Spend any time at all with Wilmot, and you'll quickly learn that there's very little down time when you're in charge of an event of this magnitude. "Mid-September through mid-November is a busy time for us," he explains. "We secure lots of our corporate sponsorships in the fall. We hit another lull between Thanksgiving and

New Year's, because it's not easy to do business over the holidays. Of course once the New Year arrives, the clubs go in the closet until after the tournament." Much of the golf Wilmot plays is of the 9 hole variety; casual outings with tournament staff members, or maybe his young son, Charlie. Because of the hand-eye requirements, every capable golfer has to be a capable athlete, but not all capable athletes are proficient in golf. Wilmot, a three-sport high school star who went on to play football at North Carolina's Guilford College, has all the potential, but little time to practice and improve. "It's kind of funny," he offers with a rueful smile. "I do pretty well in the Hilton Head recreational basketball and softball leagues, but my golf and tennis, which are the real staples of this island, leave plenty to be desired."

John Farrell and I have four things in common. We're both in our early 40s, hail originally from western Massachusetts, like to crack wise with regularity, and we each powder the golf ball a mile down the fairway. All right, so it's just three things in common.

Farrell likes to have fun on the golf course, and he provides levity and perspective in equal abundance. When I sent a wayward drive rocketing into the trees, he was quick to ask, "who's writing about your golf game today?" I wasted no time in reminding him that I was the only member of the group still at level par, a short-lived boast, as it turned out. The Harbour Town head pro is spending less time playing golf, and more time playing dad these days. There are three young Farrells at home, Megan, Thomas and Charlie, grade school age and younger. "I used to play twice a week, but now I'm lucky to play twice a month," admits John, who's been at Harbour Town for more than a decade. They say that the short game is the first to go, and Farrell illustrates. He covered 510 yards with a drive and an iron on the 515-yard 5th hole, but took three to get down from the fringe, recording a frustrating par that must have felt like a bogey.

Gary Snyder was born in Ohio, but save a two-year stint in California, has been in the south for more than 30 years. He's made Hilton Head his home for more than 20, spending six years as the superintendent at Moss Creek, a dozen at Haig Point, and now some five years at Harbour Town. He plays more golf than Wilmot and Farrell, and his smooth action through the ball and with the putter belie his mid-teens handicap. Unfortunately, he has a tendency to unleash a garish hook shot off the tee once in a while. A monster hook isn't very practical on any golf course, but on the claustrophobically narrow fairways of Harbour Town, it can be an abject disaster. Snyder showed his class with a pretty deuce on the picturesque par 3 17th, and then explained the major changes that were completed in 2001 on the island's flagship golf course. "We addressed three major issues. The first was the tee box restoration, which included laser leveling. We've also added another set of tees. Secondly, we added drain tile to all the bunkers and replaced the sand. Lastly, we restored the greens to their original specifications and contours. By replacing the Tif-dwarf putting surfaces with the new super-hybrid Tif-eagle grass, we're able to cut the greens lower, and make them putt faster." Snyder feels the golf course is in very good condition, considering it sees 40,000 rounds of play annually, and lacks a sandy soil base that facilitates easy drainage. "Our customer comment cards are much more favorable than they once were," he concludes.

The golf wasn't extraordinary, but the afternoon was, and as we all shook hands in the shadow of the lighthouse on the 18th green, I wondered when we all might get together again. Of course, Wilmot's plate is full as always. Besides the little matter of running the island's marquee event every spring, he's recently been named to a prestigious post as a board member of the PGA Tour Tournaments Association. He's also something of a newlywed, married to self-described "best friend" Dene, the first person he met on the island back in the mid-80s.

Farrell has a pro shop to run, members to service and daily fee players to accommodate. He also has to help his pharmacist wife Jane with their three small kids. Snyder has to make sure that the daily onslaught of paying customers feels that course conditions warrant an outlay of 250 bucks, while making sure they don't do too much damage to the property in the process. I imagine we'll all get together and play again soon. Maybe in 2005.

ESSAY: MARTY HOLMES AND THE HUB OF THE HERITAGE

I've got an embarrassing admission to make, one that most self-respecting golf writers would never declare publicly. I don't really love going to golf tournaments. Playing the game? Anytime, anywhere. Reading about it? I've got a den's worth of golf books, nearly all of them well read. Watching it? The Golf Channel is on the TV remote's speed dial, and I check out the weekend telecasts of any event this side of the Southern Farm Bureau Classic.

What I don't relish is traipsing around two hundred acre tracts with the milling throngs, attempting to get a viewing angle from a tee or a green, and vainly searching the horizon for a shot that probably won't be visible when it lands. Golf beats ski racing and bobsledding as spectator sports because you usually don't freeze while you're watching, but not much else. I wouldn't turn down an invitation to Fenway Park, Wrigley Field, Madison Square Garden or any venue where the action unfolds before you, but golf tournaments are invariably more appealing from the comfort of the recliner.

Of course, there are exceptions. Augusta National during Masters week is an experience that every golf fan should enjoy at least once. The inherent viewing difficulties are about the same as anywhere else, but simply strolling amidst the pines, taking in the views and munching on a two dollar sandwich wrapped in Masters green paper is worth the hassle of procuring entry.

The Heritage at Harbour Town makes for a fun day or weekend, and is in many ways the highlight of the Hilton Head calendar. The scenery is delightful, whether you're looking across the water towards Daufuskie Island or the action on the Quarterdeck. There are dozens of elite players in the field, a mellow but knowledgeable band of spectators, and a quirky delight of a golf course that's intimate, easy to walk and almost universally admired by the participants.

Everyone has a preferred viewing area at the tournament. There's always a congregation near the practice putting green, where you get an up-close look at the pros honing their strokes, and are close by the first tee and ninth green at the same time. Others sit in the shadow of the lighthouse, watching the action from tee to green on the daunting 18th. These and a dozen others are perfectly fine vantage points, but to my mind there's really only one spot to be. Befriend Marty Holmes, and grab a seat in his ivory tower overlooking the par 3 17th.

Holmes is a principle of Holmes-Smith Development, a successful corporation that develops office and industrial space, based in Columbia, about 160 miles from the island. Much more importantly, he's a fun-loving guy, a gregarious host, and has mastered the key elements of both real estate development and hospitality tents. It's all about the location.

"We've been coming to this event for more than a decade, and have had this hospitality tent location for at least the last five years. We were in the right place at the right time when this spot became available, so we grabbed it quickly," explains Holmes, showcasing the savvy intuition that has made him a business success. "We invite our clients down from all over the country, and many of our guests come year after year." Not hard to understand, considering Marty's party affords shelter from either sun or showers, comfortable seating, an endless buffet, free-flowing bar, and most importantly, an unencumbered view of perhaps the most dramatic hole on the course, the dangerous 17th.

If you tire of watching the parade of pros launch mid-irons at the flag from 190 yards, just look to either side of you. Chances are a luminary will be close by. "We've entertained the governor, Tour players and even major champions over the years, but just as importantly, the booth gives us the chance to thank our many clients and friends who've supported us over time," states Holmes.

The *Carolina Morning News* tent is located just steps away from Marty's greenside location; a prime spot overlooking the 18th tee and the water beyond. I cover plenty of ground during tournament days, from the media center, the press bullpen behind the 18th green and various hospitality tents. Our newspaper booth is great, but I find myself spending just as much time at the Holmes-Smith sky box. After all, you can always see the Calibogue, but an up close view of a Tour pro making double bogey is a rare occurrence indeed.

ESSAY: NON-PLAYING WIVES

I spend an insupportable amount of time on the golf course, and have borne witness to literally hundreds of circumstances that can best be described as contemptible.

In the last fifteen years, thousands of rounds, ten thousand hours and fifty thousand golf holes there have been innumerable highlights, but plenty of lowlights (and lowlifes) as well.

I've seen club-throwers, turf-gougers, whiners, crybabies, sandbaggers, spike-draggers, excuse-makers, cheaters and braggarts of every stripe. I've shaken my head at the foot-wedgers, mis-markers, change-jinglers, shadow-casters, cart-gunners, scorecard-fudgers, club-losers, divot-leavers, ball mark- neglecters and bunker rake-ignorers.

I've observed the freak that walks off in mid round because of poor play, the meek that walk off the course at the first raindrop or rumble of thunder, and the geek who heads home at the first cell phone call from his wife.

I've seen par 3 tee shots leaning tremulously over the edge of the cup, an eighth inch from an ace. Skulled bunker shots that hit the stick at a million miles per hour and dropped in, and four putts taken from five feet.

I've played with fat guys who had suction cups at the end of their putters, and millionaires who played with range rocks. I've witnessed family members who whiffed five times consecutively, almost losing their bladder laughing in the process, and the perpetually flatulent who cannot take a swing or mark their ball without an ad lib.

But for all the weird and wacky, the peculiar and pathetic behavior I've been privy to,

there is one thing that stands out from the rest: Non-playing wives who accompany their husbands on the golf course.

It's a particular brand of vapidity that keeps these spouses in tow, sitting endlessly in the golf cart while hubby thrashes it sideways. Perhaps I'm missing something here, but one can only use their own frame of reference for any sort of comparison.

My own tennis playing, yoga teaching, child rearing wife wouldn't deign to spend four minutes on the couch watching Tiger Woods, never mind four hours in a cart watching me stumble through the woods. That's not to say she's never ambled the linksland. As newlyweds, my fitness fanatic would shoulder the bag and caddy for me from time to time, including one particularly memorable episode in the tenth month of pregnancy when she went into false labor. But that's another story entirely.

Virtually every non-playing wife I've seen (surely there are husbands out there as well, but I've personally never encountered one) reposes like a benign Buddha.

I've met up with these non-playing wives who are "just along for the ride" from time to time. But a spate of encounters in recent months, including a listless lady who was riding shotgun for the duration of a four day golf trip, really got me thinking. It's not as if I ever had a bad experience in the company of a tagalong; none have offered unsolicited swing advice or squealed the brakes in my backswing. It's just their presence, like knowing there are Pauly Shore movies available to rent at the video store, that bothers me.

I'm all for togetherness and close, caring relationships, but aren't there fifty other more fruitful alternatives to lazing the day away in an E-Z-Go? Perhaps a movie, a tour, some shopping, a lunch date, a library or museum? Treat it like the kid's soccer practice. Drop him off, and then pick him up later. If one insists on being attendant from the opening drive to the final putt, wouldn't a book, a Gameboy, a Walkman, a Rubik's Cube or pair of knitting needles help the time to pass?

Virtually every non-playing wife I've seen (surely there are husbands out there as well, but I've personally never encountered one) reposes like a benign Buddha. Some might assist, however cursorily, in the search for a lost ball. Most are willing to drive the cart around to the back of the green, and all are willing to commiserate after the inevitable bad shot. Other than that, like the occasional one iron you see popping up in a golf bag, they seem to have no real purpose.

I was matched up recently with a struggling threesome consisting of two couples, minus a cart-warming wife. After inquiring about her non-participation, the husband told me, "She's the smart one."

I chuckled along with him, but immediately thought that his descriptive wouldn't have been my first choice. Maybe sluggish or lethargic. Perhaps insipid, or bored-to-tears. Truth be told, the word "smart" never would have occurred to me.

Course Discourse: Arthur Hills Course at Palmetto Dunes Resort

Number 15. Courtesy of Greenwood Development Corporation

Even though his course designs have played host to the PGA Tour, Champions Tour, Nationwide Tour, NCAA Championships and various USGA competitions, it seems like Arthur Hills is the Rodney Dangerfield of architects, garnering little respect. He's won a basketful of awards from the golf press, including Architect of the Year from Golfweek Magazine, but Hills' profile remains several notches below his better known contemporaries named Jones, Fazio, Dye, et al.

Perhaps his PR machine isn't what it should be, as Art Hills consistently turns out attractive, thoughtful and challenging golf experiences. A prime example is the self-named Arthur Hills Course at Palmetto Dunes.

It's evident from the first fairway that this Hills creation isn't your standard Lowcountry golf course. Players are confronted with a sizable berm fronting the opening green of

this simple par 4, a substantial mounding that camouflages almost half the flagstick. Indeed, one of the notable features of this layout is its setting among natural dunes. There's also a rolling quality to the terrain, making uneven lies the rule through many of the fairways.

This is thankfully not a basher's course, measuring a modest 6,650 yards from the extremities. It's a rare and altogether pleasant experience for an average hitter to have an opportunity to play from "the tips" without being delusional, or busting a gut trying to reach the fairway or carry hazards. The 129 slope rating is an indication that there's plenty of inherent difficulty, though the length may be quite manageable. Chief among these are lagoons threatening more than half of the holes, and tight out-of-bounds stakes lurking ominously behind certain greens.

The 2nd hole affords a perfect example. It's an average par 4 of some 375 yards, with a sizable lagoon fronting a large, odd sized green. There's room for a 270 yard tee shot, so most folks can wail away, stay short of the water and have a mid or short iron home. The problem comes if you over-club on the approach, to ensure carrying the water hazard. There are OB stakes directly behind the green, just a few feet behind the cart path. The same scenario is played out on the home hole as well. Some serious mounding will contain all but the most egregiously air-mailed shots, but those insidious white stakes are still waiting close behind the final green nevertheless.

The course truly starts to show its teeth on the 12th hole, a 400-yard dogleg right with water threatening both drive and approach. Even shots that touch down dry but tight to the waterline will likely follow the slope into the hazard. The 16th and 17th are nerve-inducing as well. Another pair of two-shot holes that are well under 400 yards in length, but again, water and wetlands must be negotiated gingerly on both the first and second shots. The course, particularly down the stretch, is a bit like a minefield in a meadow.

He's won a basketful of awards from the golf press.

Conditioning is quite good, especially considering the volume of play. The course was constructed in 1986, but the greens were redone about ten years later. The putting surfaces are excellent, offering true roll at medi-um fast speeds. The atypical layout, shot-making challenge and memorable scenery make an afternoon at the Arthur Hills Course at Palmetto Dunes a worthwhile golf experience. Little wonder it remains, more than 15 years after creation, one of the island's most popular daily fee destinations.

PERSONALITY:
PAUL HAHN JR. – TRICK SHOT LEGEND

Hilton Head resident Paul Hahn Jr. is a good guy with an engaging personality. Lucky for him, because it would be very easy to dislike someone who makes golf look so easy. We all know that the game is impossible to conquer, so how does Hahn, a grandfather in his 60s, manage to consistently produce such an array of mind-boggling shots? This amazing trick shot artist will hit balls covered with newspaper, balls that are swinging from a string, and two balls at once. He'll use a 10-foot long driver, or a driver with a rubber hose. He hits balls sitting on a stool, kneeling on the ground, or while looking back at the audience. One of his most amazing feats is swinging a five iron right handed, and then turning the club over, swinging left handed, and producing a shot of nearly identical quality. The upshot is that practically every shot is hit harder, straighter and more powerfully than most golfers could manage with both feet on the ground, using conventional technique and equipment. The audience is left to laugh with envy, shaking its collective head, and muttering "can you *believe* that last shot?"

Hahn was born in Charleston in 1940. He learned the trick shot repertoire from his late father, who began touring the world in 1948, and was an annual fixture at Masters Wednesdays. Paul Jr. describes the humble beginnings of his career. "When I started out I was doing exhibitions at high schools, Elks Clubs, VFW's and the like. Of course my dad was my mentor. He told me that if you went to a small enough town all you had to do was show up. The townsfolk would come out just to see what you were doing. I'd go through small towns in the Midwest hanging up flyers and posters that advertised the performance. Then I'd return two weeks later and do the show. We charged $2.00 a person."

From such humble beginnings are careers made, as his pay rate has multiplied by more than a thousand since he began. It's been close to 40 years since Hahn began the touring life, and he's barely slowed down in the interim. He's performed in every state in the union on numerous occasions, and in more countries than states. Trained as a pilot in the Air Force, Hahn has worn out four private planes over the decades and has accrued millions of frequent flyer miles besides. In all the travels, one particularly grueling journey stands out still. "I was hired by a major Australian newspaper to work the Australian Open in Melbourne," he recalls. "It took me more than 40 hours of travel time from Palm Beach. I did my show in front of 6,000 people, and was so worn out from the trip I actually swung and missed on one of my tougher shots. It's the only time in thousands and thousands of performances I can ever remember whiffing."

The travel demands of his profession haven't discouraged him though. South Africa, India, Singapore and Okinawa have all been part of recent itineraries.

> ## *"My problem is I'll occasionally put a rubber hose swing on my real driver."*

A logical question is, how good a player is Hahn with both feet on the ground? As a young man he attempted a career on the Caribbean Pro Tour, but it didn't pan out. "I guess I missed the discipline of the Air Force," he recalls with a smile. "I was pretty young, and I was playing as much off of the golf course as I was on it." Decades later, friends encouraged him to give the Senior Tour a shot, but he dismissed the idea out of hand. "I figured the same guys who were kicking my butt 25 years earlier would be happy to do it again. The only difference being that they now had 25 more years of competitive experience, while I had 25 less." When he's not touring, Hahn runs clinics at Palmetto Dunes Resort, and plays a legitimate round about once a week. He can still shoot near par, but

when he thinks about 'work' on the golf course the results can be disastrous. "My problem is I'll occasionally put a rubber hose swing on my real driver.

I lose my concentration for a second, and the ball is so far offline Lewis and Clark couldn't find it."

A few years back Hahn reached a career pinnacle of sorts, when he appeared as a panelist at the World Conference on Information Technology. This is a major seminar that convenes once every four years at different locations worldwide. Some of his conference colleagues included Mikhail Gorbachev, Margaret Thatcher, Michael Dell and Steve Forbes. Pretty rarified air for a guy who began by barnstorming his way through the Dakotas. Hahn, who's hit balls off of the Great Wall of China, brought his extraordinary skills to every continent save Antarctica, and could never begin to count the places he's been and the people he's befriended, will never forget that special weekend. "I've worked for presidents, Prime ministers and queens," concludes Hahn, who first came to Hilton Head in 1975. "But meeting these giants of politics and industry made me as nervous as stepping to the first tee at St. Andrews."

Course Discourse:
The Golf Club at Indigo Run

Number 17. Courtesy of The Golf Club at Indigo Run

The Golf Club at Indigo Run is pretty sweet, and it's not just because they give out homemade chocolate chip cookies in the pro shop. This private facility has a unique claim to fame. It is the first ever co-design by Jack Nicklaus and his eldest son Jackie. The course has some typical Nicklaus traits; yawning fairway bunkers, reachable par 5s with a risk factor and plenty of tee shots where a fade is the desired shot. But Jack Junior has had his share of influence as well.

The most interesting and unusual facet of the golf course is a design feature rarely associated with a Jack Nicklaus course. Almost half the holes have a collection area that borders the green. These collection areas are similar to the fringe that surrounds a typical golf green. The difference is that instead of only being a couple of feet wide, some of these collection areas are half again as large as the green itself. The result is that an approach shot that doesn't hold the putting surface will eventually come to rest as much as 30 or 40 feet away from the green. Now the player is faced with a variety of playing options, depending on their skill and comfort level. These delicate shots may be putted, chipped, pitched or flopped towards the flagstick. It is an exercise in the creative options of the short game, and it makes this course distinctive and memorable. What makes the situation even more untenable is the fact that most every green has serious undulations. Not overly speedy, but with plenty of pitch and roll. Even a skillful player could find himself on or near the putting surface in regulation, and still have plenty of work left to earn a par.

The Golf Club at Indigo Run affords the player generous landing areas on almost every tee shot. There are very few holes where a player cannot miss the fairway on one side. Just because you hit the fairway, however, does not guarantee you an easy approach shot to the green. The Nicklauses have placed their greenside bunkers ingeniously; from the point of the approach shot they don't seem intimidating, but they are ball magnets for shots that are off line to the putting surface. It isn't an especially tight course. There is virtually no out-of-bounds, and not many stands of trees or shrubs where balls will regularly disappear. It is a straightforward course, but at the same time can easily sneak up on the careless player. One can walk off a green having just made a double bogey or worse, thinking "how did I manage to mess up that hole so badly? It didn't seem all that hard."

The par three holes are very robust. They all have forced carries over marsh or water, and three of them are at least 185 yards from the middle tees. As an indication of their difficulty, two of them are rated among the hardest seven holes on the course. This is a rare attribute; normally the four par three holes on any given golf course are rated among the five or six easiest holes on the course.

One of the longer holes worth noting is the par 5 7th; at just over 500 yards it is appropriately named "gamble." A solid drive down the right center allows a player the option of going for the green in two. The dilemma is that a shot struck directly at the flag must carry a lagoon that is angling away from the player, and towards the green. A more prudent play is to attempt to carry the water further left, leaving a simple pitch towards the putting surface. But a large tree can easily wreak havoc with a shot that is played too far left. The most prudent play is a lay up shot, which should leave a short iron to the flag.

Also intriguing is the short par 4 15th. This hole is called "strategy," but might easily be re-named "baseball." If the tee box is home plate, there are large fairway bunkers where you would find first base, second base, and third base. At just 330 yards, the best play is a long iron or fairway wood played to the middle of the "infield," or fairway, just short of the middle bunker. This will leave an uphill approach of about 120-140 yards to a green that slopes from back to front.

The Golf Club at Indigo Run is not a spectacular looking course; there is no single hole, view or vista that will take your breath away. The land that the course is built on is utilitarian, without a signature river, marsh or lake to lend it memorability. The strength of the course is its conditioning, playability, and straightforward but challenging design. The members must love this place. The course can fight you on every shot during a round, but clear thinking and solid execution can easily translate to a score worth recounting.

ESSAY: TIGER WOODS TELEVISION

In retrospect, it seemed inevitable. First it was The Golf Channel, a programming idea that was met with incredulity and derision when it was first introduced in 1995. Little did the cynics realize that there was indeed a market out there for 24 hours of golf programming, be it instruction, infomercial, highlights, old matches, collectible shows, breaking news, etc. Now the television executives are gaga over the ratings that Tiger Woods pulls down. They are massive when he's in a runaway, phenomenal when he's in a dogfight and respectable even when he's in a rain delay. Factor in the solid ratings his "made for TV" matches shown in prime time the last few seasons featuring players such as David Duval and Sergio Garcia, and the next step in the evolution should come as no surprise. Get ready for the Tiger Woods Channel, whose slogan is already sending industry analysts, advertisers, and fans into a frenzy. All Tiger, all the time.

"This channel will be one of the most widely anticipated debuts in the history of television," effuses network president and former mini-tour player Richard "Dick" Dentyne. "There's nobody on the face of the earth, not an actor, rock star, politician or athlete that has captured the imagination and intrigued the populace to the degree this young man has. We expect to capture almost 10% of the market immediately, which is unprecedented, and eventually have up to 35% of all televisions in use tuned to the Tiger Channel." Programming is still being finalized at this time in anticipation of a January 1st launch, but there are a number of different programs already in production. These include:

The Prince of Thais. Pat Conroy and Kultida Woods are co-hosts in this travelogue, where viewers are taken on an extended tour of Thailand. Highlights include never-seen footage of Tiger dining on pot stickers and green curry, visiting a Buddhist temple, and touring the red light district in Bangkok.

Better 'n Eldrick. Dr. Joyce Brothers hosts this panel discussion, featuring in-depth interviews with those who have changed their names and gone on to stardom. Scheduled guests include Tony Curtis, Marilyn Manson, Eminem, Prince and Kareem Abdul-Jabbar.

Tiger Beat. A variety program geared towards teenage and pre-teen girls, all professing their love and adoration of the striped one. Wacky contest winners are rewarded with autographed photos, golf balls, and other paraphernalia.

Tele-tubbies. Craig Stadler, Tim Herron and John Daly discuss the PGA Tour, and what it will take to challenge Tiger's supremacy in the coming years.

Stanford and Sons. Co-hosts Tom Watson and John Elway profile both the famous and infamous products of the storied University in Palo Alto, where they prove continuously that graduation is no prerequisite for success.

Tiger Kwon Do. Tips and techniques on the ancient martial art that helps give the Big Cat his edge, along with the fanatical conditioning, practice habits, focus, weight training, diet and mental strength that are his other hallmarks.

Who Wants to be a Millionaire? Join caddie Steve Williams as he shares the secrets that keep him in the top 50 on the PGA Tour money list without ever having to swing a club.

Father Knows Best. Earl Woods discourses on the tough love tactics that helped mold his son into a champion. Early scheduled lectures include: "Why Tiger will be bigger than Gandhi," and "Scotland is no place for a Black man." Richard Williams acts as moderator.

Tiger's Tips. Join the phenom himself for this lively and informative instruction hour. Learn how to hit a sawed-off 230-yard 2 iron, nuke a 350-yard driver, and make any length putt through sheer force of will.

"This channel will be one of the most widely anticipated debuts in the history of television ..."

Insiders speculate that the Tiger Woods Channel might eventually spawn a host of imitators, including either the Karrie Webb Channel or Annika Sorenstam Channel. Sources within the company that refused to be identified refuted that belief however, claiming that either a test pattern or a test of the Emergency Broadcast System would likely garner better ratings.

COURSE DISCOURSE:
COUNTRY CLUB OF HILTON HEAD

Number 18. Courtesy of The Country Club of Hilton Head

Architect Rees Jones has created notable golf courses all over the country, and fortunately for Lowcountry residents, Jones has done some of his finest work in this region. Hilton Head itself is home to three Rees Jones courses, while Haig Point, widely acclaimed as one of his finest creations, lays a short boat ride away on Daufuskie Island.

The Country Club of Hilton Head is the most recent addition to this Rees Jones foursome. Located on the island's north end in Hilton Head Plantation, it has been challenging and intriguing area golfers since it opened in 1987. The Country Club of Hilton Head has a lower profile than many area courses; an understandable occurrence when you're sharing the same area code as world class venues like Harbour Town, Long Cove and Secession Club. Nevertheless, this scenic semi-private course, which stretches to almost 7,000 yards from the back tees, has long been acknowledged as one of the island's most rigorous and memorable tests of golf. The course features a variety of different types of holes, elevated greens, and ever present pot bunkering. Players must negotiate the Intracoastal Waterway, marshland and other water hazards to reach the greens safely.

The Country Club of Hilton Head is a very pretty and pastoral golf course, winding as it does through pine forest, marshland and lagoons. Five different sets of tees vary the course yardage by as much as 1,600 yards, and depending on the tees chosen, the slope rating ranges from a benign 119 up to a formidable 132. Regardless of length, every player is faced with similar difficulties. Virtually every hole on the course is a dogleg; some subtle, and some severe. The elevated greens are no trifling issue, either. In some cases the mounding is so dramatic, that an approach shot missing long or to the wrong side of the green will leave a pitch shot to a flagstick that can barely be seen.

Number 12 is a truly memorable golf hole; a sweeping, downhill par 5 playing to a length of 575 yards from the back markers.

The forward tees don't provide much in the way of driving obstacles, but in several instances, gold or even blue tee players will have to fire their ball through a tree-lined chute to find the fairway. The best example of this disparity can be found near round's end, on the 17th hole. A fairly straightforward and gentle par 4 of less than 300 yards from the whites, the hole transforms into a 390-yard gauntlet from the blues, complete with a framed tee shot that must negotiate a crossing lagoon. The signature hole is also found on the second nine. Number 12 is a truly memorable golf hole; a sweeping, downhill par 5 playing to a length of 575 yards from the back markers. A solid tee shot will land in the vicinity of the highest point on Hilton Head, and though the elevation might only be about 25 feet, (this is the Lowcountry, after all) players will be rewarded with the vista of the Intracoastal Waterway and Skull Creek lurking behind the green. This is no time for daydreaming or picture taking, because of more immediate concern is the ball-eating ravine located in front of and

slightly left of the putting surface.

The one-shot holes are notable as well. Number 7 is a mere 143 yards from the middle tees, but features a fairly steep bank leading to an ominous lagoon located on the right. The fourteenth hole requires a tee shot that must clear a wide expanse of marsh to reach the green 160 yards away. Beyond the marsh are a series of green side bunkers guarding yet another elevated putting surface. The stiffest challenge comes early in the round, at the third hole. At only 156 yards from the middle markers, length isn't the issue. This is a wide-open hole, one of the few such perspectives on the course. A sizable lagoon fronts the green, featuring a bank so steep at first glance it looks more like a wall. The successful shot will travel high and land softly on the putting surface. Any sort of line drive runs the risk of slamming into the bank and drowning, or sailing well over the green. This tee shot is a microcosm of what makes a golf course exciting and rigorous. A player must execute the required shot at the right moment, or face the score-swelling consequences.

COURSE DISCOURSE:
ROBERT CUPP COURSE AT PALMETTO HALL

Number 6 green, number 7 fairway. Courtesy of Greenwood Development Corporation

The Robert Cupp Course at Palmetto Hall Plantation is an intriguing golf course. It's about a decade old, but for some reason seems newer than that. There have been plenty of homes built there over the years, but they are pretty spread out and don't really encroach upon the course. There are lots of woods on the property, but rarely will you find the forest to be particularly dense. Most of the trees serve to frame the holes; wayward shots that find the hardwood will more often than not be playable.

The Cupp Course is probably best known for its geometric bunkering. The architect eschewed the typical bunker shapes of round, teardrop or oval; and instead constructed most of the bunkers with straight lines. You will see square bunkers, as well as rectangular and triangular ones. There are even a few that employ a zigzag pattern. But this unusual feature is mostly ammunition for the marketing department. While actually playing the golf course, most players would be hard pressed to notice much of a difference. If a first-time visitor to the facility hadn't seen a brochure with aerial photos, or wasn't told by his playing partners to notice the bunkering, I would wager at least half of the customers wouldn't perceive a difference.

This public-access course is no softy. From the back tees it stretches almost 7,100 yards, and carries a slope rating of 144. Even the middle tees are tough; better than 6,500 yards long with a slope of 136. There is plenty of water and inland marsh at the Cupp Course, and a heady player will think long and hard before pulling out the driver on certain holes. Only four of the ten par 4s are longer than 400 yards from the middle tees, and the creeping marsh, lurking water and ever-present bunkering make a three wood a viable play on several holes.

Among the more notable holes are the par 5 sixth, a dogleg left of just under 515 yards. The hole is anchored by a series of obtrusive pyramid mounds on the right portion of the fairway, some that are ten feet tall. If you can steer your second shot away from these obstacles, you are confronted with a delicate third shot that will flirt with water to the left of the green. It's a fine test.

The short eleventh hole is also well thought out. At just over 300 yards, it doesn't appear to be a difficult par 4. But the drive must carry not only wetlands, but also a series of subtle fairway undulations which could cause problems with stance and ball position. The short iron approach to a horseshoe-shaped green must carry a large bunker that is framed by sizable mounding. Bombing an approach shot up and over the mound will put the aggressive player in a lateral hazard behind the green. It is a hole that demands skill and finesse to yield par or better.

The Robert Cupp Course at Palmetto Hall Plantation is a fine choice for residents and visitors alike. The course is challenging, the setting tranquil, and conditioning above average.

PERSONALITY:
TIM MOSS – INSTRUCTOR TO THE STARS

Butch Harmon, Jim McLean, Tim Moss, John Jacobs, David Leadbetter. You might only recognize four of the five names, but that doesn't mean they don't all belong in the same sentence. Tim Moss is arguably the greatest instructor you've never heard about. He's held a number of high profile teaching positions around Hilton Head Island for more than three decades, but has remained a little known quantity to the golf world in general.

"I always considered myself one of the top 10 teaching professionals in the country," states Moss without a hint of braggadocio, and a quick inspection of his resume makes it hard to argue. "There are several reasons why certain instructors become quasi-celebrities," explains Moss, semi-retired from full-time instruction. "I've basically been a club professional, and have never taught at a resort. I've never sponsored or run a golf school, all my instruction has been done one on one. I was always a hands-on instructor. I never had an assistant or apprentice teaching lessons in my stead."

There's no such thing as a star instructor without a star pupil, as evidenced by Leadbetter's relationship with Nick Faldo, and Harmon's long-time tutelage of Tiger Woods. Moss chose a different route. He's been based in the Carolina Lowcountry since he took his first assistant's job at Harbour Town Golf Links more than 30 years ago. He's piled up a small mountain of accolades during that time, which includes becoming the 36th PGA member to earn the classification of "Master Professional." He's also been named Hilton Head Professional of the Year, Carolinas Section Professional of the Year, Carolinas Section Teacher of the Year and a two-time PGA of America Teacher of the Year nominee.

"I've known Tim for many years," states Will Mann, former president of the PGA of America. "He's very astute, a great teacher, and takes a lot of enjoyment in helping other people learn golf. He cares about the game and the people who play it, and that's what the PGA of America is all about. He also happens to be a wonderful person. He's always been a good friend to me, and a good friend to golf."

For the most part though, Moss has been below the radar. This isn't to imply he's never taken a star turn. He had a high profile job a few years back with some high profile students, a group that'll always have 'Q' ratings far larger than their handicaps. Moss served as the swing coach and golf technical advisor for the cast of *The Legend of Bagger Vance*, and worked closely with director Robert Redford, Will Smith, Charlize Theron, and most importantly, Matt Damon. "Matt was an exceptional athlete, but had no experience playing golf," recalls Moss, a man in his early 50s. "We had 28 days to turn him from a rank beginner into someone who could compete believably on film with Bobby Jones and Walter Hagen. That was far and away my most challenging task."

But not his only task. Besides spending five hours, six days a week with Damon, he had to become a PGA Tour rules expert circa 1931, the era when the movie was set. He also had to verify the authenticity of player conduct, the positions of caddies, standard bearers, announcers and galleries, and help actors Joel Gretsch and Bruce McGill replicate the mannerisms and swing characteristics of Bobby Jones and Walter Hagen, respectively.

Fortunately for Moss, he had an uncommonly deep store of knowledge about Hagen, thanks to a close friendship with PGA Tour pioneer Leo Fraser. "Leo hired me to be the head professional at Moss Creek Plantation in 1978," says Moss, describing the position he held until 1986. "Leo was an incredible guy. He was one of the first presidents of the PGA, and prior to that he was Walter Hagen's playing partner in the 1930s. He and Hagen would barnstorm around, taking on the local hotshots in best ball

matches. Needless to say, he shared many stories, recollections and memories of the Haig with me over the years. We became so close that my wife Mary and I named him Godfather to our son 'T'."

Moss' friendship with Fraser could well have tipped the scales in his favor. "I don't think the *Bagger Vance* production people were aware of my unique knowledge of Hagen when they interviewed me for the job, but it certainly didn't hurt my chances!"

Veteran character actor Bruce McGill portrayed Walter Hagen in *Bagger Vance*, a dream role for the long-time golf nut and single digit handicapper. McGill has known many golf pros over the years, and ranks Tim Moss at the top. "Besides his extensive knowledge of the game and its history, Tim has a unique teaching ability. He can observe your natural rhythm and aptitude, and help to build a repeatable and useful swing from that point. Most pros that I've observed will tend to try and break your swing down before they put you back together. Tim is different," claims McGill, who is best known for his featured role in the comedy classic *Animal House*. "He has a great eye, sees what you have that is solid and can be used in a good golf swing, and builds on those points, instead of knocking away at the faults. The trick is that over time, he replaces faults with broadening strengths that he taught or were there to begin with. It's all about positive reinforcement with him, and he never forgets that it's supposed to be fun."

Tim Moss can hit the shots, not just teach the shots, according to Kent Damon. "I was in the last foursome with Tim one afternoon during a break in the shooting of *Bagger*

Vance, recalls the star's father. "He had about 170 yards over water to the final green, which was surrounded by cast and crew. Knowing everyone was watching him as 'the golf guy,' he stuck it to five feet." The flash of golf skill was exciting, but Damon Sr. was just as impressed by Moss' personality and work ethic. "He still gets excited when he connects with a student. The enthusiasm he still shows after all his years in the business is a quality I've not seen in other instructors I've known." The Boston-based 16 handicap was pleasantly surprised at how rapidly his famous son took to the game. "Not only did his skills evolve quickly, but I also credit Tim with teaching Matt to show the proper respect for the game."

"Tim Moss can hit shots, not just teach shots." - Kent Damon

A firm believer in the game's fundamentals, Moss feels that the key to success is as simple as A-B-C. Understanding the importance of Alignment, Balance and Control is essential for any golfer. Influenced in his career by professionals such as Dr. Gary Wiren, Craig Shankland and the aforementioned Leo Fraser, Moss puts great emphasis on clear and simple communication. "To be a good teacher, you have to be a good communicator," he explains. "You may have only one methodology, but you have to be able to say things four or five different ways until the concept registers with a student."

Moss thoroughly enjoyed the Hollywood whirlwind that accompanied his involvement with *Bagger Vance*, and the attendant relationships he developed with cast and crew. Although his name appears in the movie's credits it hasn't fazed the veteran instructor, who has 30 years of PGA membership. He continues teaching a little bit, predominantly junior programs and the occasional corporate golf school. His latest project is a book of golf instruction, with a forward written by friend and former student Robert Redford. Tentatively titled "10 Ps to Awesome Golf," Moss asserts that critical elements such as posture, pre-swing, path and position are the keys to success.

"I feel I'm every bit the equal of the half dozen marquee instructors whose names you always hear," concludes Moss. "Their situations are completely different from mine, with the golf schools and being out in the public eye. As far as knowing the swing and fundamentals though, I'm just as good, if not better. My teaching career went in a different direction, but it certainly doesn't mean my students were any less important."

ESSAY: AN OPEN LETTER TO THE MEMBERSHIP OF THE HILTON HEAD AREA WOMEN'S GOLF SOCIETY

Dear Fellow HAWGS:

I'm writing today on the behalf of my dear, sweet, misunderstood cousin Lorena, who has unfortunately been pilloried by the press in recent times, in my opinion, for no good reason. I side with her not out of familial obligation, but because I could partially understand her mounting frustration with men. Bear in mind, I personally adore men. I've married three of them so far, and I've got a fourth lined up should the need arise. Neither Lorena nor myself bear any resemblance to her other sister Edwina, who was a man-hater from day one. I remember when we were just getting out of high school years ago, and Edwina confided in me that she was "looking forward to serving men." Little did I realize at the time she was planning to write a cookbook.

In any event, I'd like to nominate Lorena for membership in our club. Some of you may be skeptical about her inclusion, but let's face it. Even the most forgiving and open-minded among us have experienced at least some difficulty with men on the golf course. As I said earlier, men are great, but occasionally they give us pause on the links. Just in the past year, I've been exposed to a variety of what might mildly be termed objectionable behavior. Let me site a few examples:

There was the swarthy fellow who insisted on helping me with my swing, even going so far as to "help" me with my address position by approaching me from behind and assisting me in swinging the club. I insisted he adhere to the court mandated restraining order that requires him to maintain a 50 foot barrier from all women between the ages of 18 and 80, lest his parole be revoked. Needless to say though, my concentration was shot for the remainder of the round.

There were the middle-aged business types who by turn were ogling and then disparaging the precious girl driving the beverage cart. It was a pathetic display under any circumstances, but the fact that they could comment negatively about her cute figure when they were all spilling out of their polyester Sans-a-Belts made the situation even more intolerable.

There was the old codger I was teamed up with not long ago, a sweet enough man, but one who insisted on befouling the air with his filthy cigar. The day was clear and lovely, but the haze emanating from his choke-inducing cheroot made me feel I was stuck in rush hour on the Los Angeles freeway.

There were the two 40ish attorneys I played with late last summer, much too young to be having "senior moments," but who nonetheless forgot about the ladies tee box on 14 holes out of 18. (I counted.) Time and time again I had to beseech them to pull to the side of the path, so I could take my rightful turn on the tee. They were as apologetic as all get-out, but wouldn't you know, on the very next hole they'd be barreling off again without a second thought.

There was the overbearing business owner and his long-suffering wife I joined up with just a few weeks ago. This blusterer couldn't break 130 with a mulligan on every hole, yet he insisted on instructing both his wife and me ad nauseam. A prime example of the old adage, "those who can't do, teach."

Every joker I've ever heard who came up short on a putt, and said either "hit it Alice" or "I tripped over my skirt."

So you see men can be the source of endless disappointment, both on the golf course and beyond. I would wholeheartedly suggest that Lorena would make a respected member of our little organization, and become a force in our inter-club matches. She's gotten past all of the press hoopla that surrounded her for years, she's concentrating, and her golf game is steadily improving. She's closing in on a single digit handicap, primarily because she's hitting the ball much longer. It took her awhile, but she's finally gotten over that wicked slice.

Sincerely;
Plum Bobbit, Secretary

COURSE DISCOURSE: PLANTER'S ROW

Number 4. Courtesy of Port Royal Golf Club

Some folks might say that the Planter's Row Golf Course in Port Royal Plantation can be described to a "T," as in tight, tree-lined and treacherous. It's also a solid test of golf.

This Willard Byrd design is proof positive that good things can sometimes come in small packages. The championship tees only stretch the course to an eminently manageable 6,500 yards, the "back" tees are a paltry 6,000 yards; yet the holes seem to play significantly longer. There are two factors in play: Wet turf and heavy air can make the course seem substantially longer from either set of tees.

Also, the encroaching trees which line the fairway on many of the holes will often make a prudent player reach for a club other than the driver, leaving a correspondingly longer approach shot to the hole being played. The scorecard yardages may look trifling, but a wayward tee game will quickly make appar-ent that this golf course can play as tough as tracks that are significantly longer.

Planter's Row might be the only course on Hilton Head that is devoid of housing, save some inconspicuous condos that are posi-tioned off of the penultimate hole. It's a great marketing angle, and sounds good in the brochures, but this is not to imply that a round of golf here is a study in serenity. While it's a rare treat to not be searching in somebody's backyard azaleas for a missing ball, there are a number of parallel fairways here, and the routing of holes often leads to a green near a neighboring tee box. This course is bustling, with players visible in many directions. Exacerbating this condi-tion is the presence of the Hilton Head Air-port near a few holes towards the end of the front nine, and US 278 close to a couple of holes on the back. It's not enough that one has to contend with narrow and twisting fairways; here you're also conscious of adjacent runways and highways as well.

The outward nine is shorter and easier than the inward; the par 4 holes average about 365 yards, compared to 385 yards on the back. Without question the hardest par on the course is the 12th, a 430-yard two-shotter that features two successive lagoons guarding the green. Facing any significant headwind, the average player must crush a drive down the fairway to have any hope of negotiating the dueling water hazards and reaching the green safely. Laying up short of the green in hopes of getting up and down for par is no simple task, either. Bogeying this hole is more than acceptable; many average players would feel fortunate to do so.

As is typical of all Hilton Head resort courses, Planter's Row receives a considerable amount of play; tens of thousands of rounds annually. Under the circumstances, the golf course is in excellent shape. There are plenty of unfilled divots in the fairways, but both the short grass and rough are consistent, with almost no hard pan or patchy lies. The greens are also in fine shape; they afford a true roll, with limited contours and average speed.

Planter's Row is slightly reminiscent of Harbour Town, in that there are few sharp doglegs, but a number of subtly twisting and turning fairways which place a premium on tee shot accuracy. It's worth a visit for two reasons in particular. It's a pleasant enough course in a resort setting, and harder to score on than an initial glance at the scorecard might indicate.

COURSE DISCOURSE: SHIPYARD PLANTATION

Brigantine Number 7. Courtesy of Shipyard Golf Club

George Cobb isn't the most famous architect on Hilton Head Island, but he might well be the most prolific. Cobb's golf course designs and co-designs are sprinkled all over the island, particularly on the southern end. A classic example is Shipyard Golf Club, in the Shipyard Plantation.

Shipyard is a 27 hole complex; the first 18 holes, Galleon and Clipper, were laid out by Cobb in the late '60s. Willard Byrd added the third nine, known as Brigantine, in the early '80s. Cobb's original 18 holes are reminiscent of the Ocean Course, in nearby Sea Pines Resort. The Ocean Course was the very first golf course on the island, and like Shipyard, wanders through the housing development,

making liberal use of lagoons, greenside bunkers, and out-of-bounds stakes.

The fairways at Shipyard have more curves than a Victoria's Secret catalog. Discounting the quartet of one shot holes, only a couple of the remaining 14 holes give you a straight-on look from tee to green. Most every hole bends left or right, daring a player to shape the proper shot off of the tee. The tee boxes were rearranged a few years back, and the blue tees now stand where the whites once did. The new whites squeeze the course to a shade under 6,100 yards, while the blues mark a more respectable length of about 6,450 yards. The back tees, stretching to a distance of almost 6,900 yards, should be avoided by anyone who can't carry a drive at least 240 yards, or plays to a handicap above single digits.

The second hole on Galleon is a classic risk/reward par 5. Only about 450 yards from the middle tees, a solid drive that clears the corner bunker on this dogleg left will afford a fairly easy approach to this uphill green. The problem is the pond that comes into play directly in front of the putting surface. This hazard forces a prudent player to lay up with their second if the tee shot isn't far enough down the fairway.

The middle stretch of holes are among the toughest on the property. If a player can make it around holes 7 through 12 within a shot or two of par, then they'll likely gain a couple of strokes on the field. The 7th is short but dangerous. It's only 345 yards, but players need to fade a drive away from the stream on the left, or better yet, hit an iron. Numbers 9 and 10 are both doglegs right. The card only reads about 380, but they seem much longer. Eleven is one of the few straight-aways, but it's 540 yards long with a tight driving area. The difficulty continues on the next hole, a 400 yard par four with a pond guarding the right side of the fairway.

The fairways are in generally fine condition, although there are plenty of gouges. It's unfortunate that so many golfers are so pressed for time that they simply cannot fill their divots. The greens can occasionally be rough depending on the time of year, but they putt better than they look. They won't win a beauty contest, but they deliver a true roll, with little bumping or hopping. George Cobb's design has at least a dozen interesting holes, and makes for a pleasant day on the links.

ESSAY:
HILTON HEAD – A "TOP 10" DESTINATION

According to the experts at Golf Digest, Hilton Head is the tenth best golf destination in the world. Now aren't you glad you are living, visiting or vacationing here?

The venerable Digest is the king of the lists. More than 35 years ago, they were the first publication to rank the top 100 courses in the world. A few years back in conjunction with their 50th anniversary, they introduced a variety of top 50 lists, including the 50 greatest players, greatest teachers and then the 50 greatest golf destinations. Granted this system is far from infallible, as witnessed by Tiger Woods' relatively low ranking of 12th all time. Of course, the player's issue was put to bed before the term "Tiger Slam" became part of the golf vernacular. If the list had run a year later, you can bet that Tiger would be riding shotgun to Nicklaus and nobody else.

It's safe to say there will be far less turnabout in the destination category, however. The ranking was determined on a 1-10 scale, with categories including variety and quality of golf, natural beauty, ease of access and other, non-golf amenities. The Monterey Peninsula was ranked first, with a score of 9.309, just five miniscule decimal places better than St. Andrews, at 9.304. It's interesting to note that we had less than a fingernail's hold on the top ten, as Hilton Head edged out Palm Springs by the thinnest margin possible, 8.402 to 8.401.

There was one real dark horse listed in the top ten, found among the more likely suspects of Pinehurst, (#3) southwest Scotland, (#5) and Phoenix (#8). Sheboygan, Wisconsin made a surprise appearance at #7, on the strength of three Pete Dye designs; the two courses at Blackwolf Run, and Whistling Straits. Our upstate neighbor Myrtle Beach was listed one slot ahead of Hilton Head at #9, and referred to as "Golftown, U.S.A." The blurb alongside the Hilton Head entry read in part, "not quite as much golf as Myrtle Beach, but not nearly as many T-shirt shops, either."

You can argue that Hilton Head is a solid top-five golf destination, perhaps even a top three.

It's a dangerous comparison, but there's a compelling argument to be made for Hilton Head appearing significantly higher on the list. The nine destinations preceding it are areas where the vast majority of courses are resort or public access oriented. Granted, Monterey would have even more of a stranglehold on the top spot if the very private Cypress Point, considered one of the top two or three courses in the world, was included in the ranking. But the other top ten destinations aren't particularly well known for their private courses. Myrtle Beach only has a couple, the U.K. entries are almost all available to the American golf public, as is Pinehurst. The Phoenix area mirrors the Lowcountry, in that there are a great variety of upscale daily fee courses in addition to an impressive number of dazzling private tracks. It has long been my contention that the finest golf in this area is behind the gates at Haig Point, Belfair, Colleton, Long Cove and the like. Factor in the private gems with Sea

Pines, Palmetto Dunes, Port Royal, the courses on Daufuskie Island and the rest, and you can argue that Hilton Head is a solid top five destination, perhaps even a top three.

What's inarguable is the fact that living in this area affords easy access to almost 20% of the destinations listed by the magazine. Besides the aforementioned Pinehurst and Myrtle Beach, you can easily drive to Sea Island, Georgia (#22), Orlando (#28), Charleston (#34), Jacksonville (#37), Lake Oconee, Georgia (#38), and the North Carolina mountains (#44). It's just like the realtor said when you first started touring this island. It's all about the location.

COURSE DISCOURSE: LONG COVE CLUB

Harbour Town Golf Links is the most famous golf course on Hilton Head, and rightfully so. The greats of golf do battle there every April at The Heritage, one of the pre-eminent events on the PGA Tour. However there are some golfers who feel that Harbour Town is not the best course on the island, and others who feel that it isn't even the best Pete Dye-designed course on Hilton Head.

Long Cove Club, in Long Cove, has a reputation as one of the prolific designer's masterpiece works. No faint praise, considering that in addition to Harbour Town, Dye has designed a number of world-class courses. Casa de Campo, PGA West, the Ocean Course at Kiawah Island, Crooked Stick, Blackwolf Run, the Honors Course and the Stadium Course at TPC Sawgrass among them. Noted architect and architectural critic Tom Doak worked on the construction crew more than 20 years ago, and has long been a

fan of Long Cove. "I thought from the beginning it would be one of Dye's best, and I love the result," claims Doak, who's most recent claim to fame is stunning Pacific Dunes. This sister course to Bandon Dunes on the Oregon coast was recently ranked as the #2 modern course in America by Golfweek. "The holes at Long Cove are well differentiated by the variety of terrain," he continues. "Lagoons, live oaks, salt marshes, and Pete's artificial sand hill ridge, designed to obscure the power line between the 6th and 8th hole."

For the most part, Long Cove is not a spectacular looking golf course. There aren't lots of stunning vistas or breathtaking views. What Long Cove has in abundance are tough but fair golf holes. Very strong par 5s, a couple of cute par 3s, and at least half a dozen excellent par 4s.

The two-shot holes aren't overly long, only a few are more than 400 yards from the blue tees. Length is secondary; they are all excellent tests, featuring subtle routing, imposing hazards, and clearly defined lines of play.

The third hole at Long Cove is exceptional. A muscular par 5, the combination of a dogleg left, prevailing headwind and water down the left side of the fairway make a score of 5 seem like a birdie. The fifth is exceptional also, but for different reasons. It's a par 4 that's less than 300 yards in length, but malevolent all the same. An iron off of the tee leaves a virtually blind short iron into a green that juts into water. Dye may be known for his mounding, but this hole is extreme. The fairway seems like a huge green carpet laid over Volkswagens. It's safe to say that not one of Long Cove's 600 plus members has a neutral opinion of #5. It's either revered or reviled.

The par 4 seventh is a classic. A slightly elevated tee shows you all you need to know. Love grass to the left, a ball magnet of a fairway bunker to the right, the flag straight away, protected by swales and bunkers. It's a great hole; straightforward, honest, and challenging.

The inward nine features back-to-back beauties. The thirteenth and fourteenth are where you take out the camera. Framed by Broad Creek in the background, lucky 13 is the shortest hole on the course. It's an easy par if you don't hit it over the green into the marsh. Or left of the green into the marsh, or short of the green into the marsh. The par 4 which follows is another super hole. The scorecard lists it at only 374 yards, but doesn't tell the whole story. A sweeping right to left dogleg with salt marsh down the entire left side and behind the green makes par

here a great score.

Long Cove features very large, fast and undulating greens, some with as many as three different levels. Players who end up on the wrong side of the flag will occasionally be lucky to come away with a three putt.

The course gets tremendous play, well over 30,000 rounds per year. Many private facilities of the same caliber do less than half that number of rounds. It is a testament to the skill of the head superintendent and his staff that the course is in excellent shape. They are meticulous about conditioning. The fairways can be dry, but carts are still forbidden to leave the cart path in certain seasons.

The facility also benefits by an abnormally high percentage of walking golfers. Almost 40 per cent of the play is ambulatory, a statistic as remarkable as it is delightful here in the land of "cart-ball." It is also gratifying to see so few unfilled divots and unattended ball marks. The membership cares not only about golf, but also about their golf course.

Long Cove Club is all that a golf course should be. It's scenic, tough but fair, and in excellent condition. Consistently ranked within the top half of Golfweek's annual listing of America's Best 100 Modern Courses, Long Cove's reputation goes far beyond its status as one of the best courses in the Lowcountry. It is undoubtedly one of the finer courses in the entire country as well.

ESSAY:
THE BUDDHA OF THE BERMUDA (A BEDTIME STORY)

The grandchildren were too excited to sleep after Tiger Woods won an unprecedented tenth Green Jacket at Augusta at the age of 48. It was pitch dark, but 9-year-old Hogan wanted to chip wiffle balls around the yard. Annika and Karrie, the 5-year-old twins, were fighting over the remote to find highlights of the historic win to watch over and over. Their little brother Arnie was just 2, but took his cues from his siblings, and was swinging his cut-down driver dangerously close to Ballybunion, the family cat.

"You better calm them down Dad," said my daughter. "You're the one who made them golf-crazed in the first place." She was right, of course.

I said, "Come on, kids. Up to bed and I'll tell your favorite story."

"The Boss of the Moss?" yelped Hogan breathlessly. "Is that when you tied the famous golfer in a tournament?" said the twins.

"Bus-a-de-muss, bus-a-de-muss," babbled Arnie, toddling up the stairs.

"Many years ago," I began as the kids settled around me, "when your mom was just a few years older than Hogan here, I played in a tournament with one of the finest players of his time, and one of the best putters in golf history."

"Loren Roberts, the Boss of the Moss!" exclaimed Hogan confidently.

I laughed in agreement, and complimented him on his fine memory.

"Was he as good as Tiger, Grandpa Joel?" asked Annika. I explained that nobody was as good as Tiger, but Roberts was a class act, a gentleman, and that like me, he was a late bloomer of sorts. Six of his eight career wins came after the age of 40, including the tor-

nament course record at Harbour Town in 1996. It was the best late career record since their older brother's namesake Ben Hogan was in his prime in the middle of the previous century.

"It was my first time playing a full round with a Tour pro," I told them as the years fell away, a far off gleam coming to my eye. "We were all nervous, of course. You don't want to embarrass yourself in front of a player of his stature. But I somehow hit my approach on the first green to about 12 feet under the hole, and rolled it in for a birdie three. The Boss has stiffed his second to a foot and a half. After he tapped in, we walked off the green and I said something silly like, "After one hole of play, Zuckerman and Roberts are tied at one under."

"Have you always been a smart-aleck, Grandpa Joel?" asked Hogan earnestly. I admitted as much.

"Have you always been a smart-aleck, Grandpa Joel?"

The children were rapt as I continued, shaking their heads in commiseration when I told them I followed up with a quick triple, and then played the next three holes three over as well. "But then lightning struck again, kids. I stiffed another iron shot to 5 feet and made the putt. I did it again a few holes later, and finished the front nine with more birdies than Roberts himself!"

All the putts were fairly short, the kind you'd hang your head about if you missed. But they weren't tap-ins either, particularly on lightning fast, Tour quality greens. It was a once-in-a-lifetime chance to have the breaks read by a maestro, and with a quick glance Roberts gave me the proper line. They all fell in. "I guess the highlight was on the par 3 twelfth hole. I managed to hit it a dozen feet above the hole, and with my heart pounding, watch it drift ever closer down the hill."

"Boy, that would've been great to make an ace, Grandpa!" yelled Hogan. It would have been too good to be true, but when I rolled home a fourth birdie, this one from five feet beneath the hole, it was almost as good.

Alas, this was a bedtime story and not a fairy tale, and the carriage was bound to become a pumpkin yet again. "Now remember kids. I tied Mr. Roberts in birdies, as we each had four. But he didn't make the doubles and triples I did, nor did he fall apart on the final holes like me, dunking balls both left and right."

"So what, grandpa, you still played good," reminded Karrie, yawning.

I shut off the light and started walking out of the room, but paused to reminisce for a moment more. "It's true, kids. He told me he couldn't recall an amateur partner making four birdies in his 20 years on Tour."

I kissed them all and knowing they were too young to understand anyway, I concluded. "There have been some great alliterative nicknames in sports history. The Splendid Splinter, Neon Deion, Wilt the Stilt and the Galloping Ghost among them. Loren Roberts was and always will be known as the Boss of the Moss, one of the best names ever. But when I rolled in a 20-foot downhill bogey putt on the last to partially salvage a disappointing finish, at least for that one morning I was the Buddha of the Bermuda."

COURSE DISCOURSE: DOLPHIN HEAD

Number 18. Courtesy of Dolphin Head Golf Club

Dolphin Head Golf Club is in many ways reminiscent of its designer, Gary Player. Like the nine time major champion himself, Dolphin Head at first glance looks like it won't be much of a challenge. But attempting to play to one's handicap there, much like besting Player in his prime in an important tournament, proves to be an arduous task.

The South African-born architect is not large is stature. Likewise, Dolphin Head, his first design effort on Hilton Head Island, is also unimposing physically. It measures a fraction under 6,600 yards from the back tees, and just less than 6,200 yards from the forward tees. What makes this course a workout is its lack of forgiveness off of the fairway and around the greens. It's one of golf's basic tenets: Hit fairways and greens, and you will usually score well.

At Dolphin Head, if you miss fairways and

greens then you probably won't. It's as simple as that. The rough thrives in the warm weather, growing thick and wiry. It's imperative to keep the ball in play, a tee shot that drifts off of the short grass will leave an exponentially harder approach shot to the green. It's a frustrating experience to be well within striking distance, no more than 170 yards from the target, and be unable to put enough clubhead on the ball to reach. Similarly, balls that must be chipped onto the putting surface are a guessing game. Prudent players will forgo delicate chips, and make sure their ball exits the punishing rough no matter what.

This private course in Hilton Head Plantation has a tough set of par 3s; averaging better than 190 yards each from the back tees. The most interesting is the shortest, the 174-yard 4th hole. This devil requires a water carry, and is guarded on the right side by an imposing tree. If the pin is cut left, then the player must carry more of the hazard. If it's cut right, then one must negotiate the tree. A player hopes to find the flagstick planted squarely in the middle of the green.

The par 4 holes are a study in contrast. There are a total of ten, and eight of them average less than 365 yards from the back tees. However the other two holes are the hardest on the golf course; the sixth and the fifteenth, both more than 430 yards. Six is a dogleg left, guarded at the corner by a series of tall trees. Its back nine counterpart bends slightly to the right, and is menaced by a bunker at the corner. If the course is soft and the air is heavy, conditions which are far from unusual, then relatively few players will have opportunity to reach the greens in regulation play. The cerebral golfer will play these holes as par 5s; be happy with a bogey, and maybe get it up and down for a scrambling par.

Dolphin Head is now more than 25 years old, and is a mature looking golf course. The trees are tall, and on the front nine especially, tightly define the landing areas. It is a pretty course, if not overly scenic. Open water is visible on a lone hole on the inward nine, reminding a golfer that they are playing on an island, after all. This is a golf course that offers a day-to-day challenge for its mostly older membership. Generally speaking, length is not as imperative as accuracy. If you are a golfer who can keep the ball in play, then you'll find Dolphin Head to be a pleasant, if not overly memorable day's diversion.

ESSAY:
AN INFOMERCIAL WE HOPE TO NEVER SEE

Late one night, (very late) on the Golf Channel....

The Following Paid Presentation is Brought to You by "THE GREEN MACHINE."

GREEN: Hi folks, I'm Hiram Green. Back when I played the Tour in the early '60s I was known as "One Putt." I never won on Tour, although I did manage top 10s in both Pensacola and Fort Wayne in '63. Even though me and the missus weren't cut out for life on the road, I developed a reputation among my peers as one of the finest putters of my generation. Skeptical? Well don't take it from me. Listen to three-time Tour winner Fred Underbuck, whose career was cut short by gout back in '68.

UNDERBUCK: Yes sir, nobody rolled 'em like old "One Putt" Green. Casper was smooth and Arnie made 'em when they counted, but Hiram had the silkiest stroke of all. Too bad his full swing looked like a man swatting at a bug in a broom closet.

GREEN: Haha, thank you Fred. It's true, my ball striking was never the equal of my putting, but I'm here to sell you, I mean tell you, that great putting is a learned skill, not something you're born with. The Green Machine is the end result of thousands of hours of practice, experimentation and refinements I've developed in my long teaching career. I've helped countless golfers who've sought my counsel at the Shlecksenvania Country Club in the Allegheny Mountains, where I've been the Head Professional since 1970, and I can help you too.

VOICEOVER: *The Green Machine combines laser technology, pendulum action, dominant eye hypothesis, even shoulder progression, the exclusive "click and stick" theory and Hiram Green's patented "wrist brace" to help you make more putts, lengthier putts, putts from all angles, putts you never thought you could make!*

GREEN: Yes, I can show you, the average, below average, or downright piss poor putter how to make sweet music on the moss. Listen to Sheila Flagwort, a five time club champion at Shlecksenvania.

FLAGWORT: Before I started working with Hiram and The Green Machine I was just an awful putter. I'd be on in two, then off in three, 'cause I'd bang the ball so hard it would trundle right off the green. Finally I got so fed up with my husband smart-mouthing me, saying things like, "are you gonna walk, or take the cart to that one?" I decided I needed to develop a better touch with the putter. "One Putt" Green and The Green Machine did the trick.

VOICEOVER: *The Green Machine comes complete with a 32 page instructional booklet, a 60 minute video tape, a foolproof deck of large illustration flashcards and custom tailored Velcro straps for the patented "wrist brace" technology that distinguishes this learning program from the inferior products on the market.*

GREEN: It amuses me to see all the new-fangled techniques these pros experiment with nowadays. Pathetic band-aids like the claw, left hand low, reverse overlap, profuse underbite, belly putters and the like. With my simple easy-to-learn system and just a few minutes practice a week, you'll be breaking the hearts and taking the cash from your weekend foursome. Don't take my word for it though. Just ask TV celebrity Nipsey Roosevelt, who you'll remember from his recurring role on the hit show "Sanford and Son."

ROOSEVELT: At Bel-Air and Riviera they used to call me "Yipsey," not Nipsey, but that was before I hooked up with Hiram and his fantastic Green Machine. I practically played knock-hockey on those tough five-footers, wobbling it up short of the hole, then smacking it past, then missing it again coming back. "One Putt" Green showed me the way though, and now I'm fleecing guys like Jack Wagner, Joe Pesci, and plenty of others who still are actually finding work in the industry.

VOCIEOVER: *The Green Machine will take 3-5 strokes off your game guaranteed, or you pay nothing other than shipping and handling. Copyrighted techniques like 'think it and sink it', 'stroke it, don't poke it' and 'control the roll, control the hole' will turn you into the type of putting demon you never dared dream you could be!*

GREEN: So folks, I urge you, don't wait another minute. Are you tired of horseshoe putts that go around the rim and come right back to you mockingly? Power lip-outs from four feet that end up seven feet from the hole? Do you twitch, and only manage to get 30 footers halfway to the cup? Do you close your eyes and hold your breath on an eight foot downhill putt with 18 inches of break? Do you concede every putt less than 3 feet in the hopes your opponent will do the same? Do you quiver over supposed tap-ins? Do your tentative rolls veer away at the last moment, or stop dead in the jaws with a piece of the ball hanging over the rim? Do you break or change putters all season long? Then let me help you! Call for The Green Machine today!

VOICEOVER: *The Green Machine is yours for only 3 easy payments of $24.99. The first fifty callers will also receive Hiram Green's exclusive audiotape covering the psychological side of putting entitled "The 10 foot tap-in." This package not sold in stores, everything's included to start you on the road to great putting (8 AAA batteries not included) so call today. Money back guarantee if unsatisfied, but expect to save strokes on the green and take the cash from your pals!*
Call today! 1-800-663-7888. That's 1-800-ONE-PUTT. Shipping and handling: $15.95

COURSE DISCOURSE:
OCEAN COURSE AT SEA PINES

Number 15. Courtesy of Sea Pines Resort

If you were to stop ten people on a Hilton Head street and ask them about golf in Sea Pines, it's a fair bet that at least nine of them will mention Harbour Town Golf Links first. This Pete Dye masterpiece is the best known course on the island, and well deserving of its accolades. But the Ocean Course, located in the resort's east end, also merits serious consideration.

John Richardson is the Head Professional at the Ocean Course, and the adjacent Sea Marsh Course. He describes Ocean as "the oldest and one of newest golf courses on the island." Originally designed by George Cobb in 1960, the Ocean Course was the very first course on Hilton Head. PGA Tour pro Mark McCumber came aboard in 1995, and completely redesigned the golf course, keeping only the original Cobb routing.

"The biggest difference is the degree of dif-

ficulty," claims Richardson. "The golf course has more forced carries than before, be it bunkers or water. The original design had more generous landing areas, with the majority of trouble on the periphery, instead of directly in the line of play." The Ocean Course resembles many modern designs in that hazards most often must be negotiated through the air. There are only three or four holes that don't require a carry over lagoon or sand. The option of running the ball up onto the green is virtually nonexistent. It makes for a continuous challenge, but will prove frustrating for a less skilled player who has trouble getting the ball airborne.

Sage advice is to leave enough time for a proper warm-up before heading to the first tee. Many courses open with an easy hole or two before a golfer is confronted with a serious challenge. Not here. At just under 360 yards from the blue tees, the par 4 opening hole isn't very long, but has water in play off of either side of the fairway. The second hole, another par 4 that's slightly over 380 yards, has water and then OB right, while bunkers guard the landing area to the left. When the starter calls you to the tee, be focused and ready, or you might well start off with a couple of bogeys or worse.

Many courses open with an easy hole or two before a golfer is confronted with a serious challenge. Not here.

After the opening travails, the Ocean Course widens and softens, offering sizable fairways in which to land a tee shot, and receptive greens. Players will post lower scores by being sure to take plenty of club for approach shots. Landing short of the putting surfaces will almost always result in a ball that's wet or bunkered, but there is very little trouble to be found behind the greens. Attempting to hit it past the pin will take most of the trouble out of play.

Unfortunately, the Ocean Course only lives up to its name on one hole, the par 3 15th. This is a solid one shot hole of 190 yards that will usually play into the prevailing breeze off of the ocean. By this juncture in the round, one might either be overheating, or tending to a scorecard that's beyond repair. In either case, a prudent course of action might be to continue on behind the green, and walk about 30 yards to the beach. There, you can take a quick dip, and forget about your golf woes.

This golf course concludes much as it begins; with a couple of holes that can leave you reeling. The 17th is the longest par 4 on the course, 430 yards, with water everywhere. The par 5 finisher is almost 540 yards long, with water all the way down the right side. If the wind is against you, then a bogey on either of them, particularly 17, is an acceptable score.

The course is in generally good condition, with putting surfaces that aren't overly fast, but produce a solid roll. The Ocean Course is a golf experience both attractive and arduous. It may not be Harbour Town, but its well worth a visit nonetheless.

PERSONALITY:
A PERSPECTIVE ON PAYNE STEWART

It's been years since Payne Stewart's fatal plane crash, but the loss still resonates today. Perhaps because his passing came so quickly after two of his greatest triumphs, the 1999 U.S. Open Championship at Pinehurst, and the shocking last-day comeback at the Brookline Ryder Cup which occurred just weeks before his death.

Unfortunately, the first golfer to win back to back Tour titles at Harbour Town is just the latest in a long line of accomplished sportsmen who fell victim to air disasters. Some of these athletes were lost at the height of their careers, while others, like Stewart, were also beginning to fulfill their potential as men, beyond their chosen arena. Here are three examples with certain similarities.

At first glance, Stewart's untimely passing most closely mirrors another golfer's life cut short more than 35 years ago. In 1966, "Champagne Tony" Lema also lost his life in an airplane crash, along with his wife Betty. Lema, though a decade younger than the 42-year-old Stewart at the time of his death, was also an accomplished player and Major champion. Both Lema and Stewart possessed envi-

able golf swings; free flowing, natural and fluid. Both players had a flamboyant streak. Stewart's distinctive attire made him one of the most recognizable players in the game, while "Champagne Tony" derived his sobriquet after promising the beer-swilling press corps he would treat them to champagne after his inaugural Tour victory in 1962. Stewart was saddled with the nickname "Avis" by many of his Tour brethren, referring to his propensity for finishing in second place so regularly. Lema was no stranger to the runner-up spot himself, finishing 2nd to Nicklaus in his Masters debut, and leading the British Open at the halfway point in 1965 before faltering. Both players left an indelible mark in the record book though, as Stewart captured three Major titles, and Lema was crowned champion at the 1964 British Open.

New York Yankee's catcher Thurman Munson was 32 when he died at the controls of his own plane in 1979. Munson, like Stewart, was a fierce competitor and excitable man, someone who fed off of the emotions of the crowd. Such were Munson's leadership qualities that he was the first Yankee captain to be elected since the immortal Lou Gehrig, decades earlier. Payne Stewart's ascension to Ryder Cup captaincy was simply a matter of time. Ironically, he perished on the day that long-time colleague Curtis Strange was named the next American captain for the 2002 matches at the Belfry. Munson and Stewart both reached the highest echelons in their respective careers. The catcher matched the golfer's two U.S. Open titles with a pair of World Series rings, captured with the Reggie Jackson-era Yankees of the late '70s. Stewart won a PGA Championship as well, while Munson collected an MVP trophy and Rookie of the Year.

The great Roberto Clemente is the man with whom Payne Stewart had the most in common. He was the closest in age to Stewart, perishing at the age of 38 on the last day of 1972. It's arguable that the Pittsburgh Pirate star was more accomplished on the diamond than Stewart was on the golf course. His two World Series titles, MVP award and twelve consecutive Gold Gloves are merely the statistical underpinnings of his exquisite abilities and Hall of Fame career. But it is Clemente the man, not the ballplayer, whom Payne Stewart most closely resembles. You might recall he died en route from his native Puerto Rico to Nicaragua, where he was bringing relief supplies to earthquake victims. The finest compliment you can pay Clemente is that his humanitarianism off the field was the equal of his grace and skill between the foul lines. There are hospitals, schools and playgrounds all over the world named after Clemente; honors bestowed that have nothing to do with

his cannon arm or discerning batting eye.

The PGA Tour prides itself on its many charitable endeavors, and few current players exemplified this ideal better than Payne Stewart. Since his passing, much has been made of the fact that he once donated his entire winner's check of $108,000 at the 1987 Bay Hill Classic to the Florida Hospital Circle of Friends, in honor of his dad, who had died two years prior. What isn't well known is that Stewart earned just slightly more than $500,000 that entire season, so his charitable contribution amounted to more than 20% of his year's pay. For that matter, Stewart's gift was a significant percentage of what he had earned in his entire career to that point, a career that commenced in 1981. He consistently gave his time, his efforts, his influence and his money to causes close to him. Just weeks before his death, he donated a half million dollars to his family's church. He spent the day after his '99 Open victory playing in a charity Pro-am at the request of his caddie Mike Hicks, a native of North Carolina.

Payne Stewart, the impeccably dressed champion with the syrupy swing, will be missed by golf fans for a long time. Payne Stewart the man will be missed much longer than that.

SECTION II:
DAUFUSKIE ISLAND

COURSE DISCOURSE: BLOODY POINT

Number 13 green, number 10 in background. Courtesy of Daufuskie Island Club & Resort

Even though it's just a short ferry ride away, golf on Daufuskie Island offers a totally different feel than Hilton Head. No news there, particularly. Staten Island is but a short ferry from Manhattan, and nobody would confuse them either.

Bloody Point is one of the three courses on Daufuskie. It's atypical for several rea-

sons. There are just a couple of homes, almost no cart paths and just a scattering of people. It's in direct contrast to your average Hilton Head golf experience, where the foursomes practically line up and take a number like at the bakery, and almost every tee shot must be steered away from someone's backyard or patio.

Bloody Point is a Tom Weiskopf-Jay Morrish design, the only course this prolific partnership ever created in the Lowcountry. That is in and of itself a bonus, as the area is overrun with far too many Nicklaus, Dye, Cobb and Fazio courses. In certain ways it's nice to have a just a single example of an architect's work, even if said architects are among the best of their generation.

Bloody Point is a parkland style course, unfortunately making little use of the water that surrounds the island. It would've been nice to see a golf course created that made liberal use of the spectacular water views that should be available in abundance. Instead the designers crafted a perfectly serviceable and playable layout that makes liberal use of wetlands. The course will challenge but not overwhelm, playing just under 6500 yards from the middle tees, with a slope rating of 128.

The front nine is a bit nondescript, its holes playing back, forth and back again in a routing with little imagination. The inward nine is where the golf course shows more character. The 10th in a tiny par 4 that can be reached with a big drive, although the prudent play is probably a mid-iron. The 13th is an excellent hole; a 400-yard par 4 with water bordering the green. A robust tee shot will call for a mid-iron to the green, but anything short or crooked, particularly a ball that ends up bunkered on the left portion of the fairway, will give the thoughtful player pause.

The golf course concludes in strong fashion. The 16th is a narrow par 4 of 415 yards, and the 17th a 175-yard par 3 with the Mungen River visible behind the green, truly the only real water view available on the property. The final hole plays back away from the water, a reachable par 5 of 515 yards, featuring a green with a bunker set in the middle. A bit odd, particularly if a player is forced to pitch over said hazard to reach a flag on the other portion of the putting surface.

It's a treat to play a golf course in such a natural state. Bloody Point is not de-conditioned in the least, as the fairways are lush and the greens roll true. Rather, the course eschews the manicured look that's so prevalent in most resort courses, and features thick rough, coarse sand in the bunkers, and a minimum of signage, cart paths and man-made interference of any kind. Players feel as they're off on a rustic golf adventure, instead of being shepherded around via a series of signs, ropes and one way asphalt paths.

It's a treat to play a golf course in such a natural state.

Green fees, which include cart usage, unlimited range balls and the ferry ride from Hilton Head Island, are quite reasonable. What's less reasonable is the $2.50 charge in the grill room for a can of soda. Granted, Daufuskie is an island, and all goods must be freighted over, but it's not Tahiti. The thick stuff bordering the fairways is penal, but there's no good reason to lower the height of the rough. Just lower the price of a Pepsi and everyone will go home satisfied.

ESSAY:
RANDOM MUSINGS ON A ROTTEN GAME

All is right with the current state of my game. The handicap and swing are both stable. With football season commencing, it's nice to see the numbers on the scorecard look more like field goals than touchdowns. But I'm not so shortsighted, not after all these years, to forget that at its essence, golf is an exercise in depravity. It's a soul-sucking pastime, an endeavor that will give you a little but take a lot, and always leave you wanting more in the end.

The humiliation of the game goes far beyond our tortured posture and indifferent ball-striking. There's more to loathe than our inability to hold up under pressure or execute a simple bunker extraction. There are indignities both large and small, and they all add up to make golf the most frustrating and most fruitless game of them all. To wit, a random sampling:

No yardage markers on par 3s. Most everyone who plays regularly has an idea of how to pace off yardage. Why do some courses not take the time to put a distance marker on a par 3, so you can add or subtract the yardage to the hole? Are we supposed to simply trust the number on the scorecard? What if the card is out-of-date, or the blue markers have been moved to the gold tee box? Give us a fighting chance, and put a plate in the ground!

Stinky ball-washers. Is it really that hard to change the water in these contraptions every few months? I've used more than my fair share that reminded me of my least favorite cologne, Eau de Savannah River.

Pins that are incorrectly replaced. Granted, getting the ball in the hole is tough, but the pin as well? I've followed many a foursome

that left the flag listing at 70 degrees a dozen times a round. Too busy rushing back to the cart to write down a seven that was probably an eight to bother replacing the pin properly, I guess.

> *Why do some courses not take the time to put a distance marker on a par 3, so you can add or subtract the yardage to the hole? Are we supposed to simply trust the number on the scorecard?*

GPS Systems. This makes an increasingly lazy game even lazier. It's probably become more popular in recent years because it combines two of many folks' favorite activities; playing golf and watching TV. Get rid of these monstrosities! I can enter my score with a pencil, there's no need for a touch screen. I'll pace off my own yardage without the assistance of some faulty satellite, order a burger at the halfway house in person, and figure out my shot strategy without benefit of the useless "pro tips" that appear intermittently. Know what GPS stands for? Garbage, Pure and Simple.

Non-bettors. I'm no candidate for Gambler's Anonymous, but I'm always willing to stake a wager on my ability to (eventually) hole out. What's with these guys that "only play for fun?" Can't risk a dollar, a drachma, a deutsche mark or a Dr. Pepper during a friendly match? There's a name for guys like you. It starts with 'C-H', ends with 'P', and isn't "chump." Well, actually chump isn't bad, either.

Aggressive bettors. As they say in newspaper movies, "stop the presses!" It's one thing to lay a wager or two, even substantial ones. But what about the guys who have to get even on #9 or #18 though, and start betting all out of proportion to the original stakes? Play the game as it was outlined on the first tee, and take your winnings or losses at the end without any desperation gambits. The money wagered always seems to come full circle eventually, anyway.

Ball hawkers. It's one thing to fish out a decent ball that's clearly visible from the shoreline. It's another thing entirely to hold up the golf course so you can rake a muddy lagoon, looking for castoffs, x-outs, and other discarded orbs that no amount of cleaning will ever restore past the color of dull parchment. Want to rake something? Rake your damn footprints out of the bunker!

COURSE DISCOURSE: HAIG POINT

Calibogue, number 8. Courtesy of Haig Point

Daufuskie Island's Haig Point Club is one of the area's finest, as noted architect Rees Jones has fashioned a routing that offers tremendous variety. "The ocean setting that we had to work with was spectacular," concedes the architect. "We had the canvas to create a very special golf course. The site features high sandy soil, magnificent vegetation, and views of the ocean and waterways. Whenever you have the opportunity to have land meet water, it creates a special golf experience."

Haig Point has inland holes and ocean holes, there are forced carries required off of some tees, others with very narrow landing areas, and some that are wide open. However, the most notable facet of this course is in the variety of holes that a golfer can choose to play. Jones built a championship caliber course called The Calibogue, a true dazzler that is over 6,700 yards from the middle tees.

But he designed a 20-hole course, not a standard 18. Jones realized that there would be plenty of members who would lack the interest or ability to play The Calibogue, with its many forced carries. So he added an extra par 3 on each nine. The additional holes were shorter, and eliminated shots that required a long carry over a hazard. In addition to the two extra holes, Jones also built alternative tee boxes on four other holes.

These separate teeing areas were also designed to eliminate much of the inherent difficulty in the hole, and in some cases change the length and character of the hole dramatically. For example, the 9th hole on Calibogue is a solid par 5 of almost 515 yards, featuring a forced carry of about 175 yards off of the tee. If one is playing the softer Haig Course, then the hole is converted to a par 4 of about 350 yards, and the carry off of the tee is eliminated. "The routing of the course in and out from the high ground to the interior to the marshes and outer banks was the key to creating a world class course," the architect explains. "The ideal routing allowed us to take full advantage of the seaside elements of the island in building the championship Calibogue Course, while at the same time, constructing the Haig Course; a more playable course for members of all abilities."

> *"Whenever you have the opportunity to have land meet water, it creates a special golf experience."*
> *– Rees Jones*

All in all, the difference in length between the Haig and Calibogue is slightly less than 300 yards, but The Calibogue is a much more difficult course. In addition to the original design of 20 holes, Jones later added an inland nine called Osprey. The newest nine holes has two beautiful par 3s over water, and concludes with a pair of monsters; a 577 yard par 5 and a par 4 of just under 440 yards. After a full day of golf, most everyone would look forward to a visit to the 19th hole. Of course at Haig Point, the bar is appropriately deemed the 30th hole.

By turn carved through forest and hard by the water, The Calibogue is an exceptionally satisfying course, both visually and strategically. The opening holes place a premium on tee shot accuracy. Cut through stands of oak and magnolias, the first four holes can intim-idate a wayward driver. Balls that find the fairway can be converted to pars or better, and the approach areas near the greens are quite wide open. The first encounter with the water comes at the 165 yard fifth. The ocean is behind the green, with the Harbour Town lighthouse in view from the putting surface. It's a nice hole, but merely an appetizer compared to the 8th hole, a majestic par 3 of 173 yards built parallel to the beach. This hole is a forced carry over marsh, with little room for inaccuracy. Short, long or right of the green spells trouble.

The back nine follows a similar pattern; inland holes to begin, and then a return to the water towards the middle of the nine. The brawny and beautiful 549 yard 14th hole is where this transition begins. This par 5 begins among the trees, narrows as you progress, and has a green set within the marsh and in view of the ocean. It's a superlative hole: dramatic and difficult. The par 3 that follows is noteworthy as well. The tee box is just steps from the beach, the distance is a shade less than 200 yards, the wind will likely be gusting, and there is trouble everywhere. Enjoy the view, but be sure and finish the swing.

Daufuskie Island is a special place, serene and lovely. There are two other fine golf courses on the island, but neither can match the splendor of Haig Point. It is an extraordinary golf course, not only the jewel of Daufuskie, but one of the very best courses in all of the Lowcountry.

PERSONALITY: ALLEN DOYLE

Allen Doyle felt almost no outside pressure at the onset of his rookie year, and for good reason. "If you went to the range and looked at all the guys practicing, I would have been identified as the one with the least chance of making it. I was older, had an unorthodox swing, and by turning pro as late as I did, it practically ensured the fact I wouldn't be successful." In January of 1996 he was 47, the oldest rookie in history, and had been a pro golfer for all of nine months. To him the PGA Tour was a training ground, not a battleground or proving ground. He figured he had a couple of years to practice hard, compete and prepare for the Senior Tour, his realistic shot at the brass ring.

"It was a great time for me," recalls Doyle, a Massachusetts native who's called Georgia home for more than 30 years. "When I turned pro in March of 1995 I had no place to play. Fast forward to January though, and I'm on Tour." Doyle didn't exactly come out of nowhere, as he compiled a stellar amateur career featuring 15 assorted state titles in Georgia, and appearances in the Walker Cup and World Team Amateur Championship,

where he teamed with guys named Duval, Mickleson, Leonard, Herron and Woods. Still, his rise from golf course manager to bona fide Tour player in less than a year is astounding. Of course anyone can get from the range in LaGrange to the Elysian Fairways of the PGA Tour in five easy steps. Just follow the Rules according to Doyle:

Declare yourself a pro.

Wangle a sponsor's exemption to a NIKE Tour event. (The Pensacola Classic, first week of April 1995)

Finish in the top 25, thereby exempting you to the following week's event.

Win the next time out, (Mississippi Gulf Coast Classic) thereby exempting you on the NIKE Tour for the rest of the year.

Win twice more, finish second on the money list, and receive full big league playing privileges the following year.

Needless to say, Doyle wasn't your typical anonymous rookie. "My first event was at Tucson, and an ESPN camera crew comes down to the range, wanting an interview. I said, you sure you want to talk to me? They had just walked right by guys like Steve Elkington, Payne Stewart and a bunch of the big boys. It caught me off-guard, and it was just then I truly realized what a different story I was out there." Most every player was aware of the older new guy, and many went out of their way to welcome him. "Both Ben Crenshaw and Steve Elkington made a point to congratulate me on the previous year, and I was laughing to myself, thinking these guys won the Masters and the PGA Championship last season, but they're telling me how good I did!"

Not everyone was backslapping the former hockey star with the quirky slap-shot swing though, as a small percentage of his new colleagues were resentful of his celebrity, and couldn't even be bothered to say hello. "The majority of guys seemed thrilled for me, that a guy who had toiled as a scrub his whole life had made it out there." Doyle estimates that he was the subject of 20 newspaper features in the 28 different cities he competed in that year, and played in a number of prestigious invitational events like The Memorial, Colonial and MCI Classic on Hilton Head, where he missed the cut in his only appearance. "My moniker as oldest rookie in history definitely helped me in this regard. I had always been different on the golf course, it didn't just begin in 1996. It had never been a benefit though, until I went on Tour."

Doyle had a few anxious rookie moments that were unique to his situation, as most first year players don't have teenagers at home. "The only tee time I ever missed in my life came in Westchester that June. It was the weekend of my daughter Erin's high school graduation, but she encouraged me to play. In the middle of the night my alarm clock went off, and I guess while I was fiddling with it in the dark I turned the clock back two hours." For some reason Doyle never checked his wristwatch in the morning for the real time, arrived late, got disqualified, and ended up back in LaGrange in time to hear "Pomp and Circumstance." "I had felt bad I was going to miss the graduation, but I ended up attending anyway. Whether it was fate or something else, I wouldn't question it." Dad's newfound celebrity as a TV golfer was briefly the source of annoyance for younger daughter Michelle as well. "I got a phone call from her teacher one afternoon," remembers Kate Doyle, Allen's wife of more than 30 years. "She had seen his name or score from a West Coast tournament, and called to wonder how Allen had managed to sign the early dismissal note Michelle had brought in that day."

Doyle scratched out a living that first year, making about $136,000, and was back on Tour in '97 after a successful jaunt through Q-School. In '98 he was in limbo, losing his regular status and waiting to turn 50 in July, although he did have the unique distinction of recording top tens on the Nike, PGA and Senior Tour in a single four-month span. "I tried to enjoy myself out there as a rookie, win lose or draw," concludes the seven-time Senior Tour winner. "I might have been an oddity, but it wasn't like I won a lottery to get out there, or was the one token recreational player chosen to play among the elite. I earned my way out on Tour, just like everybody else."

COURSE DISCOURSE: MELROSE

The Jack Nicklaus-designed Melrose Course on Daufuskie Island is an untamed beauty. This golf course has many sides to its personality. It is lengthy, and tough. It's a bit shaggy, and a little rough around the edges. More than occasionally it is drop-dead gorgeous.

This course takes the combination of location and designer, and distills them down to their essence. Daufuskie is an island with few people and many trees. Nicklaus was a brazen, intrepid champion, whose game and talent were overpowering. A Nicklaus-designed golf course on the island of Daufuskie is a memorable combination.

The middle tee markers at Melrose are a shade less than 6,700 yards long, and play to a slope rating of 130. Although a 130 slope is daunting in its own right, Melrose seems even tougher than that. There is some combination of water or weeds fronting at least

ten tee boxes. On the few occasions where there are none, players will usually find serpentine fairway bunkers waiting to grab a drifting drive.

Many of the sand areas play as waste bunkers, and not as conventional traps. These hazards contain very hard, gritty sand, and a player may ground his club. While potentially easier to exit than a traditional bunker, hitting an accurate approach to these elevated greens is no easy task.

Melrose has a woodsy feel, similar in some ways to courses in New England. If you ignore the Spanish moss then many of the inland holes are reminiscent of wilderness courses in Vermont and northern Massachusetts. There are no parallel fairways at this facility, so a foursome plays in relative isolation. It is tranquil and serene.

The 522-yard, par 5 third hole is classic

Nicklaus. For the overwhelming majority who have no hope of reaching in two shots, the hole presents an interesting choice. After a decent drive, you can hit a little lay-up shot left, and then cross a pond to the green with your third, about 150 yards away. Or you can thread the needle to the right; hit between trees on the right and the pond left, and if successful, have a short third to the green with no water in play. With either choice, you are going to pay the piper. The question is when?

The course concludes in spectacular fashion. Jack hasn't been this good, this late in a round, since his back nine charge at the '86 Masters.

The ninth is just brutal. The drive is a carry of about 175 yards. From the safety of the fairway, a well-struck wood or long iron will end up within 130-170 yards of the green, provided it hasn't been pulled into the marsh, or pushed into the trees. Now the approach shot is again over marsh, and cannot drift right. It's enough to make a golfer take up tennis.

The course concludes in spectacular fashion. Jack hasn't been this good, this late in a round, since his back nine charge at the '86 Masters. Number 17 is a superb par 4 of just over 370 yards. Framed by woods left and a series of windblown trees on the right, this hole practically borders the beach, and is the best on the course. It doglegs slightly right,

and the green nearly juts into the ocean. Hit it, and putt for birdie. Miss it long or right, and live in oblivion. The final hole is a beauty of a par 5, also parallel to the beach. Psychopaths or scratch players can attempt to fly their second shot over the ocean to the green. Mortals will play left, wedge on in three, and walk away content with par, or perhaps birdie.

Green fees include boat transportation from Hilton Head, and are well worth the seasonal tariff charged. Melrose is a wild and lovely ride, one of the most underrated courses in the Nicklaus oeuvre.

Essay: My Favorite Club

My driver comes out of the bag repeatedly, sometimes to the detriment of my score. Unless the course is absolutely pint-sized, I'll unleash it at least a dozen times, often 14, and if there's a sizeable enough par 3, as many as 15 times a round. For that matter, if I'm swinging well and feeling reckless, I'll launch it off the fairway if the mood strikes. The putter is no bench warmer either; like most greens-challenged amateurs, I'm rolling the ball anywhere between 28 and 35 times a round. While the blade and the billy club are far and away the most used, and by a certain definition the most indispensable clubs in the bag, they're not my favorite. That distinction belongs to my 8 iron.

I should note here that by most standard definitions, I'm a fairly strong recreational golfer. I maintain a single digit handicap, while intermittently shooting scores in the 70s, occasionally within a few shots of par. I don't use my 8 iron in the way that most folks do, however. There are occasions when I'll take a long, full swing with my utilitarian favorite, attempting to launch the ball greenward from 120-130 yards. Most of the time though, I'm only half that distance or less from the pin, and that's when my graphite shafted Hawkeye changes from a mere club to a weapon.

I love to play golf in the United Kingdom and Ireland because my favorite golf shot is that British Isles favorite, the bump and run, or as I usually refer to it, "batting" the ball down the fairway. Keep your pitching wedge and your sand wedge; I'll leave it to those with more finesse and timing to master the 60-degree wedge, lob wedge and attack wedge. Just give me my choked down eight iron, de-lofted, played off the back foot, and "batted" towards its final destination.

My golf pals are generally contemptuous of my plodding technique, derisively referring to me as "Batman." The truth is it's a name I've grown rather fond of, especially since I'm Robin them of the cash in most every nassau or skins game we play.

Slapping the ball towards the pin has its limitations, I realize. Unless I develop a fuller arsenal of shots, there'll be no Publinx, Mid-Am, or for that matter, club championship in my future. Likewise, the occasional pond, marsh or bunker-fronted green forces my hand, in which case I grab the wedge, take a deep breath, and attempt to loft the ball to safety. The simple fact is I miss a hell of a lot more greens then I hit, and am confronted with a variety of approach and recovery shots a dozen times a round. Given a clear path to the pin, my choice is perfectly clear.

For some reason, I harbor an irrational fear that the loft of the pitching or sand wedge is too steep, and I'll either stick the club in the ground, or scream it over the green into God knows what. Sweeping the ball off of the sweet spot of my eight, honed after years of practice, occasionally leaves me a tap-in, usually leaves a legitimate shot at par, and rarely results in a score worse than bogey.

"Nonsense," my buddies say. "It's all in your head. There's so little difference between the loft of an eight iron and a wedge. You're not going to stick the club in the ground; it's all psychological."

"Maybe so," I reply. "But so is golf."

SECTION III:
BLUFFTON

COURSE DISCOURSE: THE JACK NICKLAUS COURSE AT COLLETON RIVER

Number 17. Courtesy of Colleton River Plantation

Colleton River Plantation's Nicklaus Course is a remarkable golf experience. It is superior in almost every important category. It is well-conditioned, wonderfully scenic, has many memorable holes, and requires sound strategy and execution to be played well. But the single most impressive attribute of this golf course can be summed up in one word.

Drainage. Drainage on a golf course is kind of like electricity in your home. You take it for granted, unless you don't have it. On two separate occasions I have played the Nicklaus Course after torrential rains. Not garden variety rainstorms, but El Niño induced downpours that caused other courses in the area to close for the day and dry out. Although there was water in some of the fairway bunkers as deep as six inches, the fairways themselves were mostly dry. Pumps were being used to remove rain water from the bunkers so they would resemble sand traps and not lagoons, yet the golf ball never plugged in the grass. Whether it's a combination of engineering, architecture, or sandy soil; the ensuing results are fantastic. The course should be re-named Colander River, because it drains like a colander.

Architect Jack Nicklaus has designed some wonderful courses in the last thirty years. He began his architectural career on Hilton Head as Pete Dye's design consultant at Harbour Town. That initial foray, along with his home course of Muirfield Village near Columbus, Ohio, are easily among the best fifty in the world. Others, like Shoal Creek in Alabama, and Valhalla near Louisville, Kentucky, have been the site of PGA Championships. Colleton River is considered by Nicklaus to be among his finest creations, and rightfully so. Nicklaus the golf legend was almost as well known for his strategic intelligence as he was for his raw power. Colleton River, playing 7,000 yards from the championship tees and 6,700 from the penultimate markers, sloped at 137 and 134 respectively, shows us examples from both facets of his golfer's personality.

Compared to some of the scenic marshy vistas elsewhere on the property, the seemingly pedestrian 7th hole could be easily overlooked. An average length par 5 of 540 yards from the tips, it is a hole that must be played in three shots by most golfers. Any sort of reasonable drive leaves you with a difficult choice. Hit your second to the left portion of the split fairway, and face a short third over a large water hazard. Or try and thread the needle with your second shot to the right side of the fairway, and if successful, be rewarded with a third shot that avoids the water. This hole is reminiscent of his work on Daufuskie Island's Melrose Course, which Nicklaus designed some five years earlier than Colleton, in 1987. It is as if the architect is saying "pay me now, or pay me later."

Drainage on a golf course is kind of like electricity in your home. You take it for granted, unless you don't have it.

In contrast, the Herculean 10th hole weighs in at a shade less than 600 yards. Large serrated bunkers guard the middle of this par 5, easily catching an errant second shot. Only the longest hitters could succeed on this hole with anything less than driver, fairway wood, and medium iron. A par here is a job well done. The baker's dozen that lead you from the first tee to the 13th green are great holes. Several are really cute and many are outstanding, such as the par 4 9th. But for the most part they are solid, no-nonsense tests; nothing that's astounding. Then you get to the 14th tee, and the vistas change dramatically. Logically, one realizes that they are on a relatively small piece of acreage in the Carolina Lowcountry. But it quickly starts to feel like you are on a cross-country golf excursion.

The 14th hole is a reachable par five that doglegs sharply to the right. From the tee box it looks like a ball driven through the dogleg will clear the dunes on Cape Cod, and end up on a Massachusetts beach. The very next hole looks a bit like northern Arizona. Sand dunes peppered with love grass line both sides of the fairway, leaving an unmistakable target line. The long par 4 16th hole has such a magnificent approach shot that it practically defies description. The penultimate hole is a smallish par 3 hard by the Colleton River, and a howling wind is as much of an enemy as is the inevitable loss of concentration induced by the view. The home hole, another strong par 4, returns not only to the clubhouse, but also to the more conventional style seen earlier in the round. It's a good thing too, as your average first time visitor will at this point be experiencing sensory overload.

Playing a course like the Nicklaus effort at Colleton River is a real privilege. In a world where far too many golf courses are paint by number, this one is a modern masterpiece.

COURSE DISCOURSE:
THE PETE DYE COURSE AT COLLETON RIVER

In the pre-Pete Dye era, Colleton River offered one of the finest golf experiences in the Lowcountry, as the original 1992 Jack Nicklaus design is at least as good as any other golf course in the region. When Dye left his mark on the property in 1998, Colleton's total golf package became virtually unrivaled, challenged only by nearby Belfair Plantation or perhaps the newer Berkeley Hall for area supremacy. These superb residential communities, each offering two exceptional golf courses, are on a different level than anything else. It's like trying to choose between Faulkner, Steinbeck and Hemingway, or Mays, Mantle and Musial. You can argue over which choice is better, but it's abundantly clear that any selection has merit, and they're head and shoulders above all the rest.

The Pete Dye Course at Colleton is a study in contrasts. It can be lengthened for tourna-ment competition to 7,400 yards with a bru-tal slope rating of 149, but almost never is. The conventional black tees are at 6,900 yards with a 141 slope. The gold tees are positioned at 6,450 yards with a 137 slope. The outward nine follows a conventional routing, while the inward nine is a windswept, wide open walk beside the river itself, featuring exhilarating views of the Chechessee River and Port Royal Sound. The fairways are pristine carpets in immacu-late condition, but the rough is a particular nasty strain of club-grabbing centipede grass. The course is interesting and intimi-dating, alternately rewarding and frustrat-ing, and visually deceptive. In short, it's a classic Dye design.

The architect has made liberal use of both sand and mounding on this course, and if the wind is blowing as it often does, a round can seem as exhausting as a Revolutionary War skirmish. It's a battle of bunkers and hills.

The front nine par 3 holes are both jaw-droppers. The third, fairly lengthy at 190 yards, plays to an elevated green, has water to the left, and seemingly as many sand pockets as a golf ball has dimples. Most are small and innocuous, about the size and depth of a child's wading pool. The sixth hole, a shade under 170 yards, has only a single bunker. However, this multi-tiered, three-dimensional monster with elevated grass berms interspersed throughout the sand stretches all the way from tee to green. It's known locally as "the moonscape," and it's unforgettable. Either of these one shot holes lends a vivid impression, but considered jointly, they are a powerful one-two punch.

The architect has made liberal use of both sand and mounding on this course, and if the wind is blowing as it often does, a round can seem as exhausting as a Revolutionary War skirmish. It's a battle of bunkers and hills.

The short 14th hole is worth noting. At only 300 yards, big hitters might impulsively attempt to drive the green. Fronted by a bunker and a huge, flag-hiding mound, this probably isn't the percentage play. A fairway wood or long iron played to the right leaves a short pitch, and circumvents the mound. The problem occurs when a ball leaks too far to the right, and ends up in the marsh which looms to the right of the fairway. It's a ticklish hole, in some ways reminiscent of Long Cove's fifth, another short but diabolical Dye creation.

As can be expected, short game accuracy is an imperative. If an approach shot misses a green by more than a few yards, it can create a potentially unfair situation due to the thick greenside rough. If a chip hits the green itself, it's tough to keep the ball near the flagstick. If a player lands the ball even a few feet short of the putting surface, in the hopes of trickling onto the green, the ball will almost always stay in the greenside tangle. It ain't easy, but that's golf.

The Dye Course at Colleton is a gem, and can be described with gem-like metaphors. A player needs the precision of a jeweler to execute great shots, and the guts of a jewel thief to actually pull them off properly. If you've got enough game, then the Dye Course will be an exhilarating golf experience. It's a diamond in the (all-pervading) rough.

PERSONALITY:
STAN SMITH – TENNIS TITAN

More than thirty years ago tennis great Stan Smith was moving in two different directions simultaneously. Sea Pines visionary Charles Fraser hired the Californian to oversee the burgeoning tennis program at the nascent resort, while Smith was still competing internationally at the game's highest level. While Smith moved down to the Lowcountry, he was moving up the world ranking concurrently, and by virtue of his triumphs at both Wimbledon and the U.S. Open, maintained the rank of world #1 in both 1972 and 1973. It's been more than three decades since Smith's heyday on the hard courts, but some things haven't changed, as he still calls Hilton Head home.

The Pasadena native is in his mid 50s now and his four children are growing up and moving on, but his travel schedule is as daunting as when he was globe trotting and chasing tennis titles. He stars continuously in an ongoing remake of "Mr. Smith goes to Washington," as well as Sydney, Scotland, London, Augusta, Brookline, and wherever else duty requires him as chairman of Stan Smith Events, just one of his many business

endeavors. He was home long enough recently to join me for a round at Colleton River's Nicklaus Course, where we discussed home schooling, favorite courses, and the business of being Stan Smith.

"I'm the chairman of the company, which is based in Atlanta, and has about five full-time employees," begins Smith, explaining his main enterprise. "We help corporations entertain their clients around major sporting events. All the Grand Slam tennis tournaments, as well as the Masters, British Open, Ryder Cup, and things of that nature. We arrange the entire outing. For example, at the Sydney Olympics we were working for American Express. We arranged meals, transportation, catering, and parties. We did just about everything except for hotel accommodations, and the airline and event tickets themselves."

Smith is a tall man with a big arc, and even though he has a pronounced hitch in his swing, he powers the ball down the fairway. He rammed a 20-footer into the cup for birdie on the first hole, and then ticked off a list of his other business interests. "Stan Smith Designs is in the business of designing tennis facilities. I've been with Adidas for more than 30 years, and been the touring professional representing Sea Pines Resort for about the same amount of time. I've been with Prince Racquets for some 15 years, and been involved with the player development program at the USTA for about the same amount of time. I've also been a playing editor at Tennis Magazine for more than twenty years, and have just released a tennis instruction book entitled "Winning Doubles."

He was getting up and in from all sorts of crazy spots, hitting the pin with chip shots, and all the while expounding on the benefits of home schooling.

Smith will play golf three times a week, or not play for three months straight, depending on his travel schedule. They say the short game is the first to go, but even though the sticks were rusty after a month of inactivity in Sydney, the results were impressive. He was getting up and in from all sorts of crazy spots, hitting the pin with chip shots, and all the while expounding on the benefits of home schooling. "We decided to home-school the kids up until high school when my oldest was in fifth grade. We thought it would be best if we stayed together while I was on the road, so we went in that direction. My wife Margie was the teacher, and I was the principal."

The Smiths make a compelling argument for the do-it-yourself method, as their boys were valedictorian and salutatorian at area high schools. "I've got three kids in their twenties and a teenager at home. Ramsey is the oldest. He graduated from Duke and is now a touring tennis pro, ranked about 600th in the world. His brother Trevor is at Princeton, playing first singles. My daughter Logan is at the University of Virginia, and plays tennis there, although she's not as committed to the game as her siblings. Our youngest daughter Austin attends Hilton Head Christian Academy."

Apparently Smith relishes the role of school principal that began at home, as one of his most recent endeavors is the Stan Smith-Billy Stearns Tennis Academy on Hilton Head. The initial class had about 20 students who came to the area to further both their secular and tennis education. The school is similar to more established sports academies run by tennis guru Nick Bollettieri and golf's David Leadbetter.

As befitting someone who converted two out of three chances in Grand Slam finals into titles, Smith also saves his best golf for the grandest stages. "I've shot two scores of 72 in my life," explains the five handicap. "One time at Pebble Beach, and once at Augusta National." Other area courses he enjoys include Harbour Town, Belfair and Berkeley Hall.

Smith still competes in the 55-and-over division at the Grand Slam tennis championships, and laughingly compares the insane prize money of today to the more modest purses of the early '70s. "I won $20,000 for winning the U.S. Open, and 5,000 pounds for winning Wimbledon. Of course the exchange rate was much better then!"

Smith was adept at getting it up and down,

but he was up and down himself through much of the middle portion of the round. He almost had bookend birdies though, matching his three on the opening hole with a smooth 9 iron to kick-in distance on the penultimate 17th, and changed the subject from golf to tennis as we headed for home. "There's a perception that women's pro tennis is more popular than the men's game at this point, but television ratings and tournament attendance statistics don't necessarily bear that out. The women are an attractive group of players and are exciting to watch, but I'm not sure they've caught up to the men's game."

We took predictable bogeys on the brutal 18th, and Smith commented on the American tennis timeline that he figured in so prominently. "I was a few years younger than Arthur Ashe, and several years older than Jimmy Connors. Then came McEnroe, and of course Agassi and Sampras will be leaving the stage within the next few years. We definitely have a lack of good American players coming up right now, and most of the world's best young players are foreign. Hopefully that'll change, though. You never can tell who's going to step to the forefront."

Essay: Mr. Molasses

I played a round of golf recently in about four hours, ten minutes. Not particularly slow, unless you consider that we were a threesome. In carts. Able to drive on the fairways. On a mild, windless day. On a closed course, where there were a total of three or four groups on the premises. Without waiting on a single shot. For four hours and ten minutes.

Yes, I made a handful of birdies on this world-class course, played to my handicap in ideal conditions, but was basically miserable. All because you can only play golf as quickly as your slowest guy, and I had hooked up with a beauty. Let's call him Mr. Molasses, or M&M for short.

How exactly did M&M add a good hour to our morning on the links? Let me count the ways:

He would stand ten yards behind his ball prior to each full shot, seemingly in a trance. After an eternity bordering on 15 seconds, he'd slowly saunter forward to take his stance.

He would crouch on the greens head down, staring at his marker as if attempting to decipher Sanskrit on the head of the coin. After a shorter eternity of about 5 seconds he'd lift his chin, and only then begin to peer at his putting line.

He'd meditate over club selection as if he were choosing a wife and not a wedge. Both before and after the swing, the club was rubbed, wiped and polished, then carefully and deliberately replaced in the bag.

It was obvious the man could play a little bit. He hit a couple of booming drives and some accurate approach shots, but he was having a grade 'A' bad day. M&M stubbed several chip shots from just off the green, and he only putted halfway to the hole on a few other occasions. But it wasn't his inability to execute shots as much as his pathetically glacial pace that caused the suffering. Under most circumstances I wouldn't have tolerated it, but there were some mitigating factors. I didn't know the man, he was in an understandably foul mood, and we were paired up in a business setting. I couldn't really needle him, goof on him or egg him on in my typical fashion. I was mired in a stew of exasperation.

Had the situation been different though, if I knew him well, knew I'd never see him again or been in a less formal setting, I would have uttered a few choice words. I considered the options as the morning droned on.

Some appropriate cliches: "Get the lead out." "Daylight's burning." "Anytime now."

Pointed remarks: "What exactly is your problem?" "Wake up!" "Earth to M&M, earth to M&M."

Discreet urgings: "Think brisk." "Remember, golf is a day game." "Growing roots?"

All my proposed witticisms were internalized however, as I remained mum, took an inordinate number of deep breaths, and enjoyed the scenery, spectacular as it was. The round eventually ended, and it occurred to me that a brief review of proper pace procedures was in order. You've heard them all before I know, but they bear repeating: One practice swing only please, two at the most. If you must go into a Zen-like trance before pulling the trigger, try and do it while someone else is hitting. Use the "thwack" of their shot as your signal to rejoin your earth-bound partners. Check out your putting line while others are putting; nobody paid green fees to watch some dimwit plumb-bob all day. Three simple words, paraphrased from the late great Sam Snead: Miss 'em quick.

Please heed this advice, your foursome will thank you for it. So will the group behind, and the group behind them and the group behind them. Ignore it at your own peril, but remember this: You might be fodder for a future essay yourself someday.

COURSE DISCOURSE:
HILTON HEAD NATIONAL

Number 6. Courtesy of Hilton Head National

Hilton Head National is a golf course that's as refreshing and unexpected as a cool breeze in August. Arriving with limited expectations, I was pleasantly surprised by the course's natural beauty, challenge and condition. It was kind of like agreeing to a blind date, and assuming she'll look like Marilyn Manson. Instead, you find a girl who looks more like Marilyn Monroe.

You might wonder what inspired the initial pessimism? The answer is past experience. The Carolina Lowcountry has a number of fantastic golf courses, but a fair percentage of the public and resort facilities can most fairly be described as "pleasant." These courses all have fine individual features, but as a whole they can be blandly formulaic. The most forgettable among them will usually meander through housing developments, OB stakes on nearly every hole, with lagoons and bunkers set down as if the architect was using some sort of paint-by-numbers set. This was the mind-set coming in, and I was way off base. Expecting it to be ho-hum, I instead found this 27-hole layout to be an unexpected gem.

Gary Player designed Hilton Head National's original 18 holes in 1989. Bobby Weed added the third nine in 1998. The three nines are now referred to as the Player,

National and Weed, and they are played in all combinations. Each nine is about 3,350 yards from the back tees, and slightly over 3,000 from the middle. If it hasn't been pouring, then the fairways will actually give you bounce and roll, so the course plays slightly shorter than the advertised length.

Hilton Head National is such a great place to play because the facility offers the rarest of island sensations: Isolation.

All three nines are solid routings, but the Weed course has the single most memorable hole. Number 6 is a shimmering beauty; a driveable par 4 of less than 300 yards. Water guards the entire right side, while a bunker series on the left can capture a conservatively played tee shot. The Weed course greens feature more crowns than the House of Windsor. An accurate approach to the putting surface will yield a reasonable putt. A ball that fades away, or scurries off of the back will lead to a dilemma. A prudent play is to stick with the putter; just bang the ball up the incline, and take your chances.

Another reason Hilton Head National is such a great place to play is because the facility offers the rarest of island sensations: Isolation. Not a house, condo, duplex or patio home is in sight. The one incongruity is the gas station/convenience store near the McDonald's on US 278. Both businesses are in view from the tee of the pretty par 3 eighth hole on the Player nine. It's not really that bad, it serves as a dramatic counterpoint to all of the glorious tranquility elsewhere. Besides, if you happen to drown your ball after making a sorry swing, you're not too far away from a Happy Meal.

The National nine is probably the toughest of the three, it certainly contains the hardest hole in the complex. Number 3 is a brute par 4, stretching over 450 yards from the back tees, with water guarding the left side of the landing area. The fact that the golf professional I was enjoying the day with sliced a 2 iron to 10 feet and made birdie there is irrelevant. Most anyone else would be willing to take a bogey five, and bound happily to the next tee.

This course is on the short list of "must play" resort and public access destinations in the Lowcountry. Hilton Head National, all 300 glorious acres, is definitely a regional treasure.

PERSONALITY: BILLY ANDRADE

For most rookies trying to find their way on Tour, the first season is an endless cycle of ups and downs, emphasis on downs, but Billy Andrade distinctly remembers June 4th, 1988, as his nadir. He had missed the cut by a mile at the Kemper Open near Washington the day before, and he and his caddie decided to play at nearby Congressional, site of Ken Venturi's historic U.S. Open triumph, on Saturday. "I used a bunch of different caddies that year," recalls Andrade. "That week I hired Mike "Fluff" Cowan, who had caddied for me at Q-School the previous fall. He started the round with five consecutive 3s, including four birdies, and was beating me by five shots after five holes. I remember thinking, 'how am I going to compete out there with Norman and Kite when I'm getting crushed by my caddie?'"

Don't pity the young Andrade though, the 24 year-old rookie with a meager $30,000 in earnings and more missed cuts than makes in the months prior to being waxed by his bagman. This lack of pity doesn't come from a present day perspective either, looking back from 2003 at a respectable 16-year career,

with four wins and more than eight million reasons the Congressional confessional should be nothing more than a distant and amusing memory. Even then, Andrade had some advantages that other rookies didn't.

Nowadays the PGA Tour runs a Big Brother program, where veterans counsel newcomers on the logistics of life on Tour. Andrade broke in five years before the program began, so he never had an official mentor. Luckily, he had four. "Brad Faxon is a few years older than me and also from Rhode Island, and he was literally like a big brother to me. For some reason Jeff Sluman took a liking to me, and we ended up rooming together. I had caddied for Fuzzy Zoeller when I was 15 years old, and he remembered me and was good to me, as was Jay Haas, who offered some valuable advice."

Fax and Fuzzy were no help on the last day of March though, when Billy had a bad hair day at the Greater Greensboro Open. It was tantamount to a home game for Andrade, who had graduated from nearby Wake Forest less than a year earlier. "I had gotten a bunch of tickets for my friends and fraternity brothers, and I ended up getting disqualified after the first round. I had a rule infraction that I wasn't aware of, and the guy I was playing with waited to tell me until after I had already signed my scorecard. Then he called an official over, we discussed what had happened, and I was told it would've been a one-stroke penalty if I hadn't already signed for my score."

Andrade's philosophy is to let sleeping dogs lie, so he declines to name the fellow rookie and poor sport who blew the late whistle. Besides, he only lasted that one season anyway. "He was ten or fifteen years older than me, but I got right in his face after being disqualified. If you're a stand-up guy you'll tell your partner you think something's amiss the minute he finishes." Andrade asks with a smile, "I wonder what he's doing now?"

The erstwhile roommates had life changing experiences later that summer. Sluman was winless and Andrade was wife-less, but both situations changed in an eight-day period in August of '88. Billy got hitched to Atlanta's Jody Reedy, and the next week Sluman won the PGA Championship in Edmond, Oklahoma. "The way that worked out for both of us was pretty cool," reflects Andrade. "I was happy for Jeff. He had been knocking on the door all year." While her fiancée was a freshman on Tour, Jody was planning the wedding and completing her senior year at Wake Forest concurrently. "We fell in love immediately," states Jody, recalling they were engaged after the second stage of Q-School in '87. "But I remember when he told me he was going to be a professional golfer I said 'that's great, but what else are you going to do?' When we first started going out I had no idea

how good he was, what his potential was, or that he was one of the top amateurs in the country. I knew things would work out for us though, whether he made it on Tour or not. When you're 20 and in love, you're not thinking about how you'll afford a home or condominium. I wasn't that concerned with money, and I don't think Billy was either."

The world's best-known bagman remembers that long ago day in Washington as well. "I got off to an unbelievable start, but things got ugly for me on the back nine," recalls Fluff Cowan, who claims Andrade eventually beat him by several shots that afternoon. "Billy's a solid player who can hit all kinds of good shots. He has plenty of heart, and that casual match not withstanding, I had the feeling he wasn't going to fall by the wayside." Cowan, who spent years with Peter Jacobsen, helped Tiger Woods win the Masters and is currently teamed with Jim Furyk, has seen plenty of great golf. It's really not surprising he proved to be such an astute judge of talent.

"That first year out there is like nothing else," concludes Andrade, who has one top 10 and three top 20s to his credit in a dozen appearances at Harbour Town over the years. "It's like being thrown into a huge fishbowl, and if you can survive the fishbowl for a few years, your game will elevate by itself. You're surrounded by greatness out there, and if you can hang in, eventually you'll rise up to a similar level yourself."

COURSE DISCOURSE: OLD SOUTH

Old South affords a prime example of a trend that's become increasingly pervasive in recent years. Some of the best public golf on Hilton Head isn't actually found on Hilton Head at all. Located across from Moss Creek Plantation off of US 278, Old South is as close as you can get to the island itself and still be on the west side of the bridge. The course makes for a fine day on the links, with some scenic vistas, staunch shot values and an excellent routing plan.

This Clyde Johnston design is too short for most from the forward tees at less than 5800 yards. Most players will prefer the blue markers at 6,350 with a slope rating of 125, or the gold tees at just under 6,800, carrying a slope rating of 129. Length is a secondary issue though, as either water or wetlands intrude in some combination on virtually every hole.

The opening hole is a microcosm of the course's personality. A straightforward par 4 of only 365 yards, any tee shot that can avoid a lagoon hugging the left side of the fairway will result in a middle or short iron to the green. Steering the ball down the middle makes the hole a cakewalk, but a pulled or hooked shot will start you on the wrong foot.

The two most interesting holes are mirror image doglegs; the 7th and 16th. On the front side the hole bends hard right, while the back nine version veers hard left. Again, length isn't the key. Both holes are about 350 yards, but require substantial marsh carries of 180 yards to a tight landing area. From the safety of the fairway it's just a short iron over another marsh to the green, but both shots, particularly the tee ball, give you plenty to think about. The drama is lessened somewhat by a logical local rule that allows a penalty drop on the far side of the trouble for the multitudes that won't clear the hazard. It helps immeasurably with pace of play, but for first-timers who don't realize they're entitled to drop their ball in mid-fairway if they fail to negotiate the wetlands, that first shot will undoubtedly make for an extra tight grip on the driver.

While in certain respects Old South offers a fairly typical Lowcountry golf experience, there are some notably positive differences. Unlike Harbour Town and a dozen other higher profile public access golf facilities, there is a minimum of housing adjacent to the links. Designer Johnston didn't go overboard with green side bunkers either. Almost half of the holes are either bunker-less near the green, or have an incidental presence only. Several of the green settings, particularly towards the middle portion of the inward nine, are situated beautifully. Lastly, Johnston's routing plan is exceptional. There are very few consecutive holes that are aligned in the same direction, nor do they simply head back and forth. Instead they proceed to all points on the compass. It's a treat under any conditions, but on breezy days particularly it gives players the opportunity to experience a variety of shots with wind affecting the ball flight from all different directions.

The course can be played in combination with sister property Old Carolina, located just a few miles further west on US 278, at a discounted package rate. This arrangement surely makes for one of the best values in the area.

ESSAY: THE ROOT OF ALL EVIL

I haven't played golf since the end of October. This is an ironic turn of events, considering that the possibility of year-round play was one of the major factors in our decision to migrate south. When I departed frosty old New England for the Lowcountry, I assumed that my heretofore-standard golf schedule, featuring a Thanksgiving week finale, was a thing of the past. This year however, my pattern is still the same. It's not due to a lack of interest, or a hectic schedule, or nostalgia for my old ways. The reason I'm off the links is as simple as it is unfortunate: I'm injured.

I had the phenomenally bad luck of engaging an inch-thick tree root in a little game of "chicken." This was a mismatch from the opening bell. Tree root in a TKO. I'm not looking for a re-match by any means, but it wasn't a fair fight. This wasn't your standard tree root, lying there plainly visible. Had that been the case I might have declared an unplayable lie, and taken a much despised penalty stroke along with my drop away from the root. Or at least I would have attempted a thin little defensive shot, holding the club lightly and attempting to play my ball out laterally. Neither of these options were a consideration, because this root was as devious as they come, a real predator. It lay there perfectly camouflaged by a thin layer of autumn leaves, biding its time, patiently waiting for an innocent and unlucky victim. I was happy to have found my ball at all, in what looked like a decent lie, with a reasonable stance, and with an unimpeded line to the green. I took out my three iron and attempted to reach said green, a mere 190 yards away. To make matters worse, my golf swing is rarely compared to those of Fred Couples or Ernie Els, they of the long and languid swings, swings that look like they are taking place under water. Mine by comparison starts quickly, picks up speed in the middle and ends in a hurried lurch. Tempo isn't my strong suit. My iron came down, met that damned root and stopped dead. It was like kicking a brick wall in sandals.

My right hand vibrated with pain. I walked up the fairway in a daze, knowing full well that I had inflicted some serious damage to my dominant hand. I rode along in the cart until we reached the turn, anticipating the ice bag that would help to alleviate the initial pain. I happened to run into an orthopedist in the grill room, who was pretty sure I hadn't broken any bones, but advised me to come see him in his office later that week.

My iron came down, met that damned root and stopped dead. It was like kicking a brick wall in sandals.

Since that fateful day many weeks ago, I have been x-rayed, bone scanned, and re-x-rayed. I have used ice, heat, ultrasound, acupuncture, and something called iontophoresis. Besides the orthopedist, I have sought help from an osteopath, a masseuse, a hand surgeon and a physical therapist. Note: Don't believe anyone who tells you that acupuncture hurts a little. They are wrong. It hurts plenty!

Now almost two months later, my hand doesn't ache or throb, but it's still weak. I shake hands lefty these days, having been caught in one too many vise grips with my right. The diagnosis is soft tissue damage, and recovery time is measured in months, not weeks.

An injury like this helps to put things in perspective. In the grand scheme of things it really is only a minor setback, but it's hard to maintain composure when it's sunny and 70 degrees on Christmas Day. It isn't easy to admit in print, but since I'm hurt I hope for bad weather. Selfish, I know, but when it's 40 degrees or raining I feel better, because there would be no golf that day regardless.

When this injury has run its course I will walk to the first tee with a brand new attitude, one that will last the rest of my life, or at least until I make my first triple bogey. I will enjoy myself more on the course, and not take for granted the fact that I am healthy and able to play without pain. Fat shots or thin, slices or hooks, shots that are stiff and shots that are sculled. I will savor them all, as long as my hand and wrist hold up at the moment of impact. And one last thing: Arbor Day is now at the top of my list of least favorite holidays.

COURSE DISCOURSE: BELFAIR WEST

One of the major reasons golfers flock to the Lowcountry is to experience the beauty of our courses. Granted, there are no spectacular elevation changes or mountain vistas, and outside of Daufuskie Island, the final holes at Harbour Town and just a few other examples, ocean-side holes are at a premium as well. What we do have in abundance is marshland, ancient stands of gnarled oaks, lush love grass and sparkling lagoons. To my mind, no course exemplifies the seductive beauty of this area better than the original West Course at Belfair.

Assessing the region's finest courses is no easy task, but somebody's got to try. Haig Point is more spectacular, the Nicklaus Course at Colleton River is a better traditional test of golf, Secession is the best walking course, and Harbour Town has the legacy, but there's no prettier golf course than

the original Tom Fazio design at Belfair. The subtle beauty of Belfair begins with the ever-present scalloped, landscaped bunkering. Virtually every hole is framed by these flowing, soft-shouldered beauties, many containing shrubs, small trees and clumps of love grass. Needless to say, one's perspective changes quickly when forced to actually extricate a wayward shot from within. However, from the outside looking in, these hazards are less of a nightmare and more of a visual reverie.

The middle tees at Belfair's West Course play just over 6,600 yards, and carry a solid slope rating of 131. Fazio keeps things interesting at every turn, and offers a beautiful balance of golf holes; some delicate and some daunting. Selecting the cutest hole on the premises is not unlike trying to pick the prettiest girl at the Miss America pageant; there are plenty of viable options, and no real wrong answer. Number 6 is a definite candidate; a 360-yard par 4 with two distinct greens. A recent visit found the flagstick located at the dangerous right side green, barely the size of a tennis court, fortified by bunkers and fronted by a lagoon. It's one of the slipperiest short irons you'll hit all day. Big hitters will make birdie more often than not on the 464-yard par 5 seventh hole. The rest of us will have to lay it up, and then wedge over the water to a severely angled green.

The fairways have more levels then a parking garage, but because the rolling terrain is so steep, it's rare to encounter a severely pitched lie.

The par 5 thirteenth and short hole that follows offer the best views of the Colleton River this side of Colleton River Plantation. Just keep your head down long enough to steer it away from the waste area and the marsh, and you'll do fine. The 16th is a tiny diamond of a par 3, the penultimate hole is less than 350 yards but can easily deface the scorecard, and as you head for home, don't look for a welcome mat. At almost 440 yards and with a substantial marsh to carry, a green in regulation here is worth noting. Most folks will smack a fairway wood down the fairway, hoping a deft chip and an accurate putt will put a four on the card.

Conditioning, as is to be expected, is excellent at Belfair. The greens, mostly large and undulating, are quite speedy. The fairways have more levels than a parking garage, but because the rolling terrain is so steep, it's rare to encounter a severely pitched lie. My single quibble, so often the case here in the land of cart ball, is that the significant commute between the twelfth and thirteenth, not to mention the return trip to the par 3 sixteenth, make walking impractical at best. A shame, really. A course this lovely and serene is a pleasure under any circumstances, but can best be appreciated while ambling.

COURSE DISCOURSE: BELFAIR EAST

Number 16. Courtesy of Belfair

It's rare and pleasurable to find a golf course that succeeds on different levels. There is no shortage of modern courses that are attractive, but overly deceptive. By the same token, many solid course routings fail to captivate, or offer little in the way of natural beauty. The East Course at Belfair is an exception to this rule. It's a combination of beautiful and playable, both testing and visually arresting.

The single most memorable feature of this excellent Tom Fazio design are the troughs of love grass-peppered sand that frame many of the fairways. These frightening and omnipresent hazards serve a dual purpose. Not only do they give the course tremendous character and dramatic appeal, they also provide an unmistakable target line from the tee box. There are a number of elevated tees, and the presence of the sand troughs bordering

the landing areas make for a clearly defined target. The design effect is extremely satisfying, as most of the holes are both straightforward and spectacular.

From the middle tees, the East Course plays slightly less than 6,500 yards, with a par of 71. There are more than a dozen delightful holes on the property, the quintet of one-shotters in particular. This "Fazio Five" ranges in length from a tiny but treacherous 137 yards, to just a shade over 200 yards. The designer must be sympathetic to faders, because the majority of the trouble is found to the left of the green. On the back nine par 3s in particular, heed some sage advice. If you're going to miss the green, miss right. Like fireworks and fisticuffs, these cavernous sand hazards looming on the left are best viewed from a distance.

A Fazio trademark is the do-or-die, diminutive par 4. The East Course's 5th hole is a prime example. The yardage is listed at 312, but that's choosing to play the hole via the fairway, aiming at 12 o'clock. If a player has the fortitude and the requisite tailwind, you might choose to aim at 10 o'clock, trying to airmail the expansive water hazard, and reach a green that's about 230 or 240 yards away as the crow flies. It's a fun and demanding hole; take a deep breath and make the choice. You might soar with an eagle, or you might swim with the fish.

It's a combination of beautiful and playable, both testing and visually arresting.

The East Course features two distinct types of sand in the hazards, a choice that was dictated by the design team. Players will find either typical fluffy-type bunker sand in some spots, and a more natural, gritty variety that has a tendency towards hard pack. It's an unusual feature, and adds a degree of difficulty to a shot that most players fear regardless. A minor consideration in the scheme of things though, as Belfair's East Course is truly a delight. It takes a well deserved position alongside the West Course at Belfair, Long Cove, Haig Point, Chechessee Creek and Colleton River's Nicklaus Course as one of the finest private venues in the area.

PERSONALITY:
DUKE DELCHER – GOLF MONARCH

I'm still not sure which is more impressive; the way Duke Delcher manages his game, or the golf course that he manages it on. One thing is certain. Delcher's ability and the original Fazio course at Belfair are both exceptional. We knocked it around recently on an absolutely gorgeous winter afternoon; 70 degrees, sunny and still. People always say to me "man, do you have a great gig," and while I'm generally inclined to agree, there are factors that not everyone takes into consideration. With a handicap in the high single digits, my swing has more moving parts than a jet engine. Playing with pros or top-level amateurs can be nerve-wracking. Normally, I can keep it together somewhat, and finish within a few shots of 80. On this day, I found plenty of trouble, and the match deteriorated into a lame television reunion show. It was Duke Delcher vs. the Duke of Hazards.

Like me, Delcher moved to the Lowcountry in his late 30s. He and his wife Debbie hail originally from Bucks County Pennsylvania, and moved to Belfair when Duke began working there as a sales executive. Unlike me, it didn't take him long to make

his mark in the South Carolina golf world. The year after he arrived he was named the South Carolina Player of the Year.

"Real estate is my profession," begins Delcher, now in his mid 40s. "Golf is my night job. My success in golf has undoubtedly helped my business success in Hilton Head, though. Up in New Jersey, I was successful in real estate sales, but very few people really cared that I was an accomplished amateur golfer. The vast majority of people around here are involved in the game, and pay attention to it. Your achievements on the course are more recognized by the public."

Duke's golf achievements are many and noteworthy. His finest moment was representing the United States in the 1997 Walker Cup Matches at Quaker Ridge, in suburban New York City. The Walker Cup is like the Ryder Cup for amateurs; the top Americans oppose their counterparts from the United Kingdom. Delcher competed in three matches, and won two. Duke has also played in the U.S. Open, about a dozen U.S. Amateurs, and half a dozen U.S. Mid-Amateurs. His grandfather, who managed a working class golf club in the Philadelphia area, introduced him to the game. His grandmother ran the kitchen. "It was the country club lifestyle," explains Duke. "Except we weren't the members, we were the help." He turned professional in '77, and tried to make it as a touring pro. He had conditional status on the PGA Tour in '82 and '84, played the Asian Tour in '81 and the European Tour in '83. "I didn't distinguish myself on any tour," he recalls. "I gave up the game in '85; I only played about a dozen rounds total in the next four years. Long before that I had decided to play until I was 30. If I didn't reach the level I wanted to attain, I was going to quit."

Duke regained his amateur status in '89, so he could compete in the U.S. Amateur at Merion. "I was motivated to play because it's a wonderful course, and it's in my hometown." He made it to the round of 16, and has been a presence in the amateur golf world ever since.

Delcher's game is so efficient, it's almost boring. He doesn't kill it, but he hits it down the middle. He doesn't always hit it close to the stick, but he doesn't miss many greens, either. He birdied both par 5s on the front, to offset a pair of bogeys. He made seven pars in a row on the back, and I missed his one long birdie from the fringe when I was down in the "gunch" looking for a ball that had air-mailed the green. It was a ho-hum 71, nothing to it, on a day when he missed half a dozen 15 footers by less than an inch each time. It didn't seem to faze him. "Sometimes they go in," was all he remarked.

"Belfair is a wonderful place," continues Delcher. "I interviewed down here as a favor to a friend in New Jersey who is a charter member. I was immediately attracted to the people. The bank tellers here thank you when you withdraw money; that intrigued me. Up in Jersey, they don't say a word if you deposit a thousand, and if you take out a hundred, they look at you like you're robbing the place."

Delcher's game is so efficient, it's almost boring. He doesn't kill it, but he hits it down the middle. He doesn't always hit it close to the stick, but he doesn't miss many greens, either.

Duke's primary office is now located at Berkeley Hall, although he still works at Belfair regularly. "My day-to-day activities are mostly at Berkeley, because that's where we're actively selling property. But depending on my client's wants and needs, I'm at Belfair also. It's really both places." Delcher's done a good job being in two places at once, as he's been the top producer in development sales at both communities since his relocation.

"Duke's status as a word-class amateur certainly doesn't hurt us," states Gary Rowe, Berkeley Hall's co-managing partner. "He is a sales leader, but more importantly, his input is invaluable when it comes to hiring golf staff, course set up, conditioning, rules and so forth. He is also our main liaison to Tom Fazio and his design team."

Even with his thriving real estate career, Delcher continues to find success at the highest levels of the amateur game. Recent highlights include high finishes at both the prestigious Hugh Wilson Memorial Tournament at Merion and the Northeast Amateur at Rhode Island's Wannamoisett Country Club in 2001.

When he's not working, practicing or competing, Duke spends time with his daughter Taylor, now in grade school. She was just a baby when they arrived, but Duke saw a bright future. "I didn't know that much about the place, but I felt subconsciously it would be a great environment to raise children. It's turned out to be a great move for us, we love it here."

ESSAY: "SNAPPEE" SOCIETY

Do you know golfers whose pace of play can only be described as glacial? Are there members in your regular group who are slower than tooth decay? Are you familiar with players who are so sluggish that a bird could nest on their heads while they stand over a shot? If you answered "yes" to any of these questions, than you are an excellent candidate for membership in the new speedy play society. It's called Slowpokes Need A Prompt Player as an Example to Emulate; or **SNAPPEE**, for short.

As a SNAPPEE member, you will be responsible for making sure that golf remains a daytime pursuit. It is our members' responsibility to keep the group moving, eradicate plumb-bobbing, and do away with all dilly-dallying. There will be special consideration given to all applicants who were born in the United Kingdom. These folks are predisposed to be SNAPPEE members, as an 18-hole match lasting longer than three hours in the U.K. is a rare occurrence indeed. Speedy golf is in their genes.

Are you familiar with players who are so sluggish that a bird could nest on their heads while they stand over a shot?

Now, membership in SNAPPEE is going to be very prestigious, and a bargain to boot. For only $49.95, plus $9.95 shipping and handling, you receive a signed certificate, visor, bag tag, whistle, and commemorative stopwatch. But all is not roses. For every SNAPPEE member who's selected, there are 50 members of the Society that Nurtures an Abundance of Inert Linksters, or **SNAIL**. As a member of the moral minority, it's up to the SNAPPEES to convert the SNAILS. Here are some examples of what neophyte SNAPPEES need to be on the lookout for:

The practice-swing pariah – This miscreant isn't content with a practice swing or two. He needs six before each shot. Add 110 ball strikes to 600 or more imaginary ones, and you can see why he's barely moving by the 16th tee.

The ball hawk – She drives a Mercedes, vacations in Europe, and spends ten minutes whacking the weeds, looking for a cracked Flying Lady golf ball that was purchased used six months prior.

The storyteller – This windbag couldn't be more fascinating; just ask him. Stories of business triumphs, girls he's dated, rounds he's shot, courses he's played, miraculous shots he pulled off; a veritable fountain of information.

The statue – A quick rule of thumb. If it takes a player longer to pull the trigger on any particular shot then it does to read this sentence, then he's an offender.

Pokey Pete – This guy moves like he's underwater. He walks slowly, he talks slowly, he takes more time to throw a few blades of grass in the air then you do to pull your club and advance the ball. Remind him that golf is supposed to be exercise, and he'd be better served to simply commit, and then hit.

The space cadet – A variety of annoyances here. Walks to the wrong cart, doesn't know where the next tee is, leaves clubs, can't find his ball, forgets to replace the pin, etc. etc. Aren't there medications available for these people?

Little Miss Wishy-Washy – Is it a hard 5 iron, or a soft 4? Should I try and hit a fairway wood out of this lie? Is it a crosswind, or is it quartering? Explain gently that her game would improve if she could lose the paralysis by analysis.

The Mulligan Man – This beauty has somehow forgotten that golf is supposed to be played with a single ball. Bad drive? That's O.K., hit another. Skulled approach? No problem, just reload and try again. Get Mulligan (and his bag of "practice" balls) off of the golf course!

There are dozens of reasons to admire PGA Tour pros. Their impressive swings, scores and paychecks to name but three. But you do not want to imitate their pace of play. There are thousands of dollars riding on the outcome of each of their shots. You probably play a two dollar nassau, and will shoot between 80 and 105 whether it's in three hours or four. Make it closer to three, will you?

Legendary golfer Sam Snead said it best when he counseled a young golfer who was making his first appearance at a professional tournament. Snead said, "Miss 'em quick, son," and his advice is worthy of being expanded upon. The most companionable golfers are quick and competent. The next best thing is quick and incompetent. This is followed by slow and competent. The least companionable golfers are.....well, you get the idea.

COURSE DISCOURSE:
OLD CAROLINA

Numbers 9 and 18. Courtesy of Old Carolina Golf Club

The tree canopy marking the entrance to the Old Carolina Golf Club is reminiscent of the stately Avenue of the Oaks across the street at Belfair. Of course Belfair is the playground of the privileged few. To gain access to the luxurious lifestyle and golf riches contained therein requires a bare minimum outlay of half a million dollars, and in most cases, significantly more than that.

Old Carolina, by contrast, is a lovely daily-fee property where even high season, prime time green fees will provide enough cash back from a hundred dollar bill for a burger and a few beers in the grillroom. Afternoon rates, shoulder season and combination packages with sister course Old South can minimize the associated costs drastically, making this Clyde Johnston design one of the best values in the region.

Fine, so it's affordable, but is it good golf?

Absolutely. Built on the site of a former thoroughbred horse farm, Old Carolina's serene setting in a series of high-grounded meadows is unlike most area courses. There are a smattering of houses on the property as well as an unsightly power line, but little else distracts from the pastoral location.

Old Carolina's serene setting in a series of high-grounded meadows is unlike most area courses.

The course is less than a minute from US 278, but there's no hint of the commotion or traffic buzz that are part of the equation at other golf locales located on the same high volume corridor.

Old Carolina plays 6,800 yards from the tips, 6,400 from the blues and almost 6,100 yards from the whites, with slope ratings of 145, 135 and 127, respectively. Neighborhood ball hawkers must have a field day prowling the grounds at dusk. Water and wetlands are abundant, influencing 16 holes directly. Only on the first and tenth, simple and straightforward par 4s well under 400 yards from the blue markers, can players be assured they'll hole out with the same ball they teed up with. That is provided they don't hook (on the first) or slice (on the tenth) over the OB fence onto adjacent Buck Island Road.

Mounding on a golf course smacks of artificiality, and there's a ton of it at Old Carolina. For some strange reason, it enhances the experience here though, isolating fairways from each other, and providing individual playing corridors where your foursome seems like the only one on the course. Although length isn't necessarily a premium concern here, accuracy off the tee is. Dead straight balls will hop down the fairway and provide short iron approaches. Offline tee shots will stop dead into some of the ungainly knolls and rises flanking the fairways. Not only does it necessitate a much longer approach to the green, but often requires an awkward stance as well, with the ball resting well above or occasionally well below your feet.

While the opening nine is perfectly adequate, the rustic nature of the golf course becomes more apparent on the inward journey, particularly on hole eleven. This dogleg par 4, again quite short at 365 yards from the blues, features a major wetland down the entire right side of the landing area. This area of the course is on the western edge of the property, close to the border of Rose Hill Plantation, and is heavily forested. Here the architect uses far less mounding. The omnipresent woods and wetlands provide the feeling of quietude naturally that had to be concocted by bulldozer in the opening sequence of holes.

It's in this section of the course where the best series of holes are found in succession. The 13th is the signature hole, a 350-yard par 4 with water threatening the left side of the landing area. A long iron or fairway wood tee ball will leave a short iron approach over water to a green ringed by ten pot bunkers. The 14th is an excellent par 3 with a wetland running from tee to green, and the next is a potentially reachable par 5 of 485 yards. There are wetlands off the tee, and water pinching the lay-up shot from both sides; a lagoon to the left and a snaking creek on the right. Bombers have the distinct advantage here, going up and over all the potential trouble and aiming a wood or long iron at a moderately sloping green.

Old Carolina is a fine test, with course conditioning the equal of many private facilities. It would make for a lovely golf course stroll if not for a series of cumbersome cart path rides, necessitated by road crossings in the nascent housing developments. It may not be a walking paradise, but as area daily-fee courses go, it's an awfully nice ride.

COURSE DISCOURSE: EAGLE'S POINTE

Number 3. Courtesy of Eagle's Pointe

Davis Love III is best known around Hilton Head for his four victories at Harbour Town in the Heritage Classic. Now the Sea Island resident has added a fifth triumph here in the Lowcountry. It is the design of Eagle's Pointe Golf Club, located several miles west of the Hilton Head bridge on US 278.

This golf course has a number of positive attributes, but one word describes the single feature which stands head and shoulders above the rest. Tranquility. This is undoubtedly one of the quietest courses this side of Daufuskie Island. A round of golf here seems to take place almost entirely in the woods, particularly through most of the outward nine. The feeling of serenity is enhanced because there are few parallel fairways; it's a relatively rare occasion to view groups other than the ones directly ahead or behind. The impression of solitude is based partially on the fact that the developers have dictated that housing will ultimately encroach on only half of the golf holes.

In anticipation of heavy play, Love has provided oversize greens, many with substantial undulation. The fairways can be spotty in places, but the quality of the putting surfaces is commendable.

Eagle's Pointe is not a particularly difficult golf course. The middle tees on this par 71 are under 6,400 yards, and carry a gentle slope rating of 126. Even the back tees are less than menacing, not quite 6,800 yards with a slope of 130. Presumably, the designer himself can easily break par here using nothing but irons. The course begins in benign fashion; a simple dogleg par 4 of 375 yards, followed by a straightaway par 5 of 487 yards. For the most part, fairways are wide and generous. Rarely will a player have to maneuver the ball to one side or the other. On most holes, anything in the short grass will leave a reasonable approach to the green. An exception is the difficult 9th hole, a par 4 of just over 400 yards. A well-placed tee shot will land left of the starkly beautiful tree which guards the right-center of the fairway. A creek meanders up the right side of this hole, and any ball flirting with the branches is strictly hit or miss. Perhaps sink or swim paints a more accurate picture.

The par 3 holes aren't lengthy, but they're crafty. Averaging just over 160 yards each, size is not the problem. Only the 11th hole, playing over wetlands in the direction of US 278, clearly defines the hazard that must be negotiated. The 8th and the 15th, by contrast, have insidious little creeks that angle slightly into the line of play. A well-struck iron on these holes will afford a birdie putt. However, any substantial mishit, either a pull or slice, will potentially find the water. The holes are dangerous, because they're a bit deceptive.

In anticipation of heavy play, Love has provided oversize greens, many with substantial undulation. The fairways can be spotty in places, but the quality of the putting surfaces is commendable. It's unfortunate that golfers here can't really enjoy a walk through these woods. Like so many courses in the area, Eagle's Pointe was designed with buggies in mind. While the walk from green to tee in many instances is insignificant, there are a couple of long hauls between holes that make walking here impractical.

Davis Love III's reputation as one of the finest golfers of his generation has been cemented time and time again on Hilton Head. With Eagle's Pointe, he continues to establish himself as an architect to be reckoned with as well. This design is a pleasure to play, and a welcome addition to the daily-fee options available in the area.

PERSONALITY: DAVIS LOVE III

My day with Davis began ominously. The occasion was the much-anticipated official grand opening of Eagle's Pointe Golf Club just off of Hilton Head. Course designer Davis Love III was scheduled to attend, and the entire production; the tournament, press conference, lunch, cocktails, gift package and mementos, seemed promising. This was in direct contrast to the weather. It was an overcast day to begin with, and the rain commenced in earnest while I was driving on US 278 towards the club.

This event had been in the works for months, and while most everyone enjoys a brush with celebrity along with a free meal, the golf round, scheduled to be the focal point of the day, was in danger of being cancelled. Improbably, the rain, which had fallen relentlessly during lunch, picture taking, and the quasi-press conference, began to abate. There were dozens of folks milling around getting ready to leave, when we were suddenly informed that the driving range was open, and the tournament would begin soon after. Somewhat reluctantly and a little skeptically, I retrieved my clubs from the trunk of my car. At that moment, there was no way of knowing that this near-washout would evolve into a real blowout.

There were two memorably amusing moments that occurred during the course of the afternoon; one visual and one verbal. The sight gag happened on the driving range during warm up. The great man came out after a short while, and stationed himself at the far end of the range. Before too long a small crowd had gathered; it isn't often one has a chance to observe a world-class player up close, without encumbrance. Directly adjacent to Davis was a local TV personality, a fine enough fellow whose area market share is but a fraction of his handicap. I hit the ball O.K., but I could never stand next to a Tour pro without a) turning around and watching him, or b) removing myself to the far end of the range, to avoid suffering a comparison. Mr. TV was either blissfully unaware, or secure in his middling ability. Davis was striping rhythmic wedge shots, while his range neighbor was cutting the grass and killing gophers. It made for a remarkable contrast; kind of like watching Janet Reno strike yoga poses next to Cindy Crawford.

The second incident came about three hours later, towards the end of our round. One of my teammate's teenage sons came out to hit a few shots with our group. I asked him his name just after he sliced an approach shot towards a green side sand trap. He replied "Bunker." I repeated myself, thinking he had misunderstood the question. "I heard you," he responded cheerfully. "My name is Bunker, that's what everybody calls me."

I was shaking my head and laughing. Not only was the timing of his reply impeccable, but the kid has a name that would make novelist Dan Jenkins proud. He would fit perfectly alongside "Two Down," "Foot Wedge," and the rest of Jenkins' marvelous golf-inspired monikers.

It made for a remarkable contrast; kind of like watching Janet Reno strike yoga poses next to Cindy Crawford.

The highlight of the day was playing a couple of holes with Davis, although we weren't exactly together. On the first hole, he launched his tee shot from the back box, 40 yards behind the rest of us, and out-drove the longest hitter in our group by about 50 yards. The next hole, I managed to stay within 25 yards, and when I knocked in a meaningless four-footer for par, I was rewarded with an offhanded comment of "nice putt." In the scheme of things it's pretty minor, but on the other hand I could've picked it up or missed it. It became more meaningful a moment lat-

er, when he casually mentioned that he's probably missed a million dollars worth of "gimmes" over the years. Dollars aside, it reminded me of an even shorter putt he missed at the final hole of the '96 U.S. Open, a putt which would've gotten him into an 18-hole playoff for the championship.

The best word I can use to describe Davis Love III is smooth, and it goes far beyond the effortless power of his golf swing. Whether it's an inherited trait or been honed after years in the spotlight, he can really handle a crowd. Observing him during the course of the day, I was reminded of Nick Price's well-known comment about how being #1 is harder off the course than on it.

Davis isn't #1, but he's very near the top, and everybody wants a piece of him. A handshake, a photo, a meaningful word, a clap on the back or a comment; everyone wants an interaction. He strikes a delicate balance; no one monopolizes him, and no one gets brushed off. I don't care if he rips a 300-yard drive, or slaps one sideways into the forest, he's still an impressive guy. Not a bad course designer, either.

ESSAY: RECORD KEEPING

The experts say the first step towards curing an addiction is to admit to one, so after much soul searching and at the urging of my beleaguered family, I'm prepared to come out in public. I'm an obsessive record keeper. There, I said it.

I make this admission now because there's light at the end of the tunnel; I'm not nearly as pathetic as I was some years ago. Sure, I still keep track of every round, but only so I can enter the proper score for my handicap. Yes, I mark down where I played as well, but only so I can build towards my lifetime goal of 500 different courses. You can argue that noting golf winnings ($237 last year, $3,246 since 1995) is somewhat obsessive, but it's only for tax purposes. Before you rush to ridicule me though, let me begin at the beginning.

When I first took up the game I was as normal as the next guy, took each round in stride, relived the good shots, vented at the bad, and forgot about them both by day's end. I guess there were signs, sure. As a kid I tracked my tennis won/lost record, and in college I kept my personal stats from the Ultimate Frisbee team, but those were aberrations, I can assure you. My downfall can be traced unequivocally to the purchase of a handsome, full-sized diary titled "My Golfing Record" at an outlet store in Freeport, Maine, almost a decade ago. With room to track handicap fluctuations, tournament results, equipment purchases and fill-in-the-scorecard replicas for 55 rounds, it was like giving Robert Downey Jr. the keys to the pharmacy. Once I got started I couldn't stop.

I preserved the precious pages in my diary for rounds below 80, from the first (July 3rd

1991, rolling in a breaking 15 footer for 79) to the last page available (October 8th, 1998, a 76 with a bogey on 18.) It is my great shame to tell you it took 7 years, 3 months and 4 days to fill it; the shame emanating not from my golf incompetence, but because I actually noted the time frame on the final page! Of course I couldn't fill the diary and be done with it, that would've been too easy. Instead, it sickens me to admit as the pages dwindled to a priceless few, I actually made dozens of Xerox copies of an empty page, so I could continue my dementia at journal's end, not unlike the tracheotomy patient who insists on smoking through the hole in his throat.

Of course for every round below 80 there were a dozen or more above, and these were noted as well, not in a hole by hole fashion, but in their own obsessive, compulsive way. Most idiots who gets caught up in the game attempt to track all of the "meaningful" statistics for a while, at least until they realize how pointless the whole thing is. I'm talking fairways hit, greens in regulation, number of putts, etc. Would you believe that for several years I supplemented the usual drivel with crucial information like how long the round took, who I was paired with, what the weather was like and how many balls I used? It seems only the type of coin used as a ball marker or whether a tree, bush or bathroom served as a comfort station escaped my notice. If my wife had been paying any attention to me during those years she surely would've had me put away.

This is all easier to admit now that I've gotten the worst of my fixation behind me. It wasn't a dramatic intervention by my

friends, as none of them knew about my little secret. Instead, it was an epiphany that
occurred one day in the summer of '99.
After a brief period of flagging interest,
entering numbers only without the associated commentary on best holes and playing
conditions, I stopped cold turkey. I wasn't
drawn to re-reading of my past "triumphs,"
and I knew nobody else would be, either. If
I was bored to tears one day and thumbed
through the accumulated notes, it would
have revealed a distressing pattern. For every
round below 80 where I parred the last few
holes or made a clutch birdie on the last,
there were at least three rounds where I came
down the stretch within a few shots of par,
only to finish in a flurry of bogeys or doubles. Who needs a constant reminder of one's
own ineptitude? I get that day by day, there's
no need to maintain an archive of incompetence. I quit then and there, and have been
clean ever since.

It feels great to treat golf like everyone else
does now, as an absorbing pastime, and not
a sick preoccupation. There are still a few
hurdles to clear, but I think tracking handicap fluctuations, (81 since '91; high of 19,
low of 5) lifetime eagles, (16, last on
12/11/02) and number of different courses
played in 79 shots or less (41) doesn't make
me that peculiar. Lots of people track the
basics, don't they?

COURSE DISCOURSE: CRESCENT POINTE

Hilton Head is a top ten golf destination by almost any standard, but the plain truth is that most visitors don't set foot anywhere near the area's premiere courses. Sure, you can cough up $250 and take a spin around Harbour Town Golf Links, perpetually ranked within the world's top 100 by GOLF Magazine, but the home of the The Heritage is the exception, not the rule, in terms of public access golf. The other elite courses in the region; Long Cove, Haig Point, Colleton River, Belfair, Secession, etc. are all private. A quartet of golf buddies on vacation from Schenectady has about as much chance of getting past the gate as they do of finding a bartender on the island pouring shots out of a liter bottle of vodka. It just isn't going to happen.

That's why Arnold Palmer's Crescent Pointe Golf Club is such a boon to the area.

It's a classy facility that is well-conditioned, scenic and challenging. Best of all, access is unconditional. If you can pay, then you can play. It's a good deal that becomes an exceptional value in combination with the Davis Love III-designed Eagle's Pointe, located just a few miles further west on US 278. Package rates are available and it's money well spent, as Palmer and longtime collaborator Ed Seay have fashioned a pretty track amidst rolling terrain that offers two distinct personalities.

The front nine looks similar to many area courses. There are lagoons and the occasional homes dotting the mostly wide open terrain, although there's no housing at all on the first several holes. The back nine is much more serene, although some houses come into view toward round's end. The encroaching woods bordering tighter fairways on the inward nine are somewhat reminiscent of a couple of local Nicklaus designs; Melrose on Daufuskie Island, and the Jack and Jackie collaboration at Indigo Run.

Some are drawn by the consistently fine conditioning, others by the convenient location, and many by the name association with The King, designer Arnold Palmer.

The blue tees are just less than 6,500 yards with a respectable slope rating of 130, while the Palmer tees are a shade under 6,800, with a slope of 137. The opening hole sets the tone of the golf course. It's a relatively simple par 4 of 360 yards, but with bunkers guarding the left side of the fairway, and a rock-walled pond fronting the green. A well-placed wood and short iron will yield a birdie putt, but plenty of players will walk off the green shaking their heads, singing the double bogey blues. The second hole is one of the most striking par 3s in the Lowcountry, a substantial downhill carry over water, with ribbons of bunkering both front left and well right. The ninth is another one shot wonder, better than 200 yards of carry over wetlands to a difficult green. One of the more memorable holes on the back is the 424-yard 13th, a severe dogleg right with a Sahara expanse of sand along the right side of the fairway. A boldly struck tee shot over the corner of the bunker will result in a short iron approach to a green guarded by scalloped bunkering and a pond. The hole requires two quality shots, but it isn't often

one can fire a mid or short iron at a flag on a hole of such length.

Crescent Pointe debuted in the spring of 2000, and very quickly became a popular choice with both visitors and residents alike. Some are drawn by the consistently fine conditioning, others by the convenient location, and many by the name association with The King, designer Arnold Palmer. They come for different reasons, but few will leave disappointed in the golf experience. This course is on a short list of area public access facilities that are truly worthwhile.

PERSONALITY:
KEVIN KING – AT THE TOP OF TWO WORLDS

Kevin King isn't necessarily the best golfer in the Lowcountry, although there are few who would dispute his position in the top tier. There are other realtors on the island who sell more property than King does, but not many. What is an indisputable fact, however, is that there is no other individual in this area who can produce such a high volume in real estate transactions, while simultaneously posting such low numbers on the scorecard. King is firing from both barrels, and is in a category practically all his own.

Impressive is not a word that adequately describes Kevin King's golf resume. Gaudy is more like it. The North Carolina native was co-captain of the UNC golf team for two years, and all-ACC from '77 to '79. He was the youngest contestant at the 1977 U.S. Open, qualifying at age 19. He's played in seven U.S. Amateurs, with an appearance in the round of 16 in 1993. He's participated in six U.S. Mid-Amateurs, restricted to players 25 and older, and also made it to the round of 16. He's qualified for two MCI Classics, and was the South Carolina Player of the Year in both '88 and '93, as well as the state

amateur champion in '89. He won the state mid-amateur ten years apart, in '88 and '98. Locally, he's captured the Hilton Head Amateur on eight different occasions, and in 1997 became the first man ever to win the Amateur and the Hilton Head Open in the same year. He was also the runner-up in the South Carolina Open in 2001. Twice the club champion at Secession, three times the champion at Long Cove Club, at one time he held both courses scoring records. But when we teed it up together at the Ford Plantation south of Savannah, I was in the lead after two holes.

Impressive is not a word that adequately describes Kevin King's golf resume. Gaudy is more like it.

"Probably once a year I start off double bogey, double bogey," laughs King, a man in his mid 40s. "You were just lucky enough to witness it. If I live long enough, you can be sure I'll do it again." King continued to play indifferently, until a slightly-pulled tee shot midway through the front nine crashed resoundingly into a tree off the fairway. He turned to the group, and slowly said, "I am about to get angry." This simple utterance proved to be something of an incantation. Not only did King shoot even par from that moment forward, but his very next shot was the finest of the day. From a downhill lie in the rough, he smoked a 200 yard 4 iron underneath a tree limb, skirting a parallel water hazard, and onto the green within 10 feet of the hole. Birdie. A couple of holes later, a 200-yard 5 iron appeared to cover the flagstick on a par 3. The ball bounced long, into a thick collar. King pulled the pin, and lofted a delicate pitch that spun around the hole, coming to rest within inches. He produced the same shot out of a bunker several holes later. He didn't make a putt over 10 or 12 feet all day, but practically every putt looked like it was going to drop. It is the mark of an accomplished player; every roll on line, with enough speed to get to the hole, but with the touch to keep what remains within tap-in distance.

Kevin King may be best known for his golf prowess, but his business success isn't too far behind. He spent eleven years with Prudential Real Estate on Hilton Head, and was the company's leading producer for all eleven years. He moved over to Charter One Realty about five years ago, and immediately became one of the company's top producers. Every year, King finishes as one of the top 10 salespeople on the island, often within the top 5. "My golf career has helped my real estate career tremendously," says King. "Every door

I've had opened for me has come through golf, including my education. When I came to Hilton Head in 1983, I met a gentleman through golf who offered me my first substantial job on the island. Later, I met the top real estate producer out here, and he offered me my first real estate job. Hilton Head is such a golf Mecca; my clients tend to be avid players. They'll often refer their friends to me, and many times during the course of a round, we'll discuss the local real estate market, and start off a nice relationship."

King and his wife Lynn, also a North Carolina native, have two teenage daughters, Cameron and Victoria. The girls are as passionate about gymnastics as their father is about golf. Consequently, they've been enrolled in a gymnastics academy since they were little, and have spent the last several years away from home. They are enrolled in Aiken, South Carolina, and living with Lynn. On weekends, either the girls return home, or dad travels to see them. "I'm lucky my wife is as supportive as she is. To stay competitive, I play three or four times a week in season, less in the winter."

Kevin King is hitting on all cylinders. He's got a stellar family, a solid career and a sterling golf game. No course records fell at the Ford Plantation, but it takes more than a commonplace round for him to lose perspective. "One thing I've learned in this game," concludes King, "there are things that'll happen that you don't expect. Golf is constantly throwing you curveballs, just like life. How well you deal with them shows what kind of character you have."

COURSE DISCOURSE: HIDDEN CYPRESS

Number 18. Courtesy of Hidden Cypress Golf Club

It sounds like a cheesy advertising slogan, but in this case it's truly appropriate. If you haven't golfed at Sun City lately, you haven't golfed at Sun City. "We estimate that only about 30% of our residents were golfers prior to moving here," states Director of Golf Bob Pfeffer. Consequently their original course, Okatie Creek, is an extremely straightforward track designed with beginners in mind. This is in direct contrast to Hidden Cypress, the second Sun City course which opened several years later.

Hidden Cypress, featuring several forced carries over marshland, omnipresent bunkering and some truly innovative greens com-

plexes, is a challenging and rewarding golf experience. And for the time being anyway, particularly on the back nine, a quiet and peaceful tour through the woods.

"Okatie Creek was designed to allow players of all levels to enjoy a round of golf, while our newer course requires some true shot-making ability. There are collection areas around the greens that attract wayward approach shots, and creative bunkering that forces players to think carefully about their strategic options," continues Pfeffer. Stretching just shy of 7,000 yards from the tips, most decent players will have all they can handle from the middle men's markers at almost 6,500 yards, with a respectable slope rating of 129. The Mark McCumber routing affords players a nice rhythm, as simpler holes often follow difficult holes, allowing the golfer to catch their breath. No hole is harder than number 4, a 430-yard dogleg that requires two powerfully placed shots to reach the green. The ninth hole is shorter by 60 yards, but any drive drifting right will catch a slope and kick down into a lateral hazard, making bogey the best you can hope for.

> *The course conditions at Hidden Cypress are excellent. The fairways are close-cropped and lush, while the generous and uniquely-sloped greens are well-maintained and true.*

Construction at Sun City is a fact of life, as the community is growing continuously. At some point, Hidden Cypress will resemble the majority of area real estate courses, with housing lining the fairways, but such is not the case currently. About half of the front side holes have no homes encroaching, while the entire back nine, framed by magnificent stands of hardwood on most every fairway, is a tranquil delight. The inward nine has a pair of notable one-shot holes. The twelfth is 165 yards over a lagoon framed by lovely Baja grass, while the 17th checks in at 216 yards, with a pair of fountains and a babbling brook adding distraction on the left. Call me a traditionalist, but I feel the only place for a fountain on the golf course is between the soda machine and the restroom. The waterspout adds some pizzazz though, and more importantly, isn't serving to mask a weak hole. The 17th is long, strong and worthwhile, with or without a fountain.

The course conditions at Hidden Cypress are excellent. The fairways are close-cropped and lush, while the generous and uniquely-sloped greens are well-maintained and true. While there's at least incidental water on 14 holes, the real menace is the bunkering. On the par 5s especially, placement off of the tee is paramount. Stay out of the sandbox and you'll often make a routine par with a chance at a birdie, but playing out of the bunkers too often here will result in round-ruining bogeys on the three shot holes. With its quietude, challenge, shot making opportunities and conditioning, a trip to Sun City comes highly recommended. Hidden Cypress is a hidden gem.

COURSE DISCOURSE:
BERKELEY HALL – NORTH

Cynics might contend that the Lowcountry needs another upscale, private golf community like Liz Taylor needs another bridal shower. But that argument would likely precede a visit to Berkeley Hall, one of the newest and in some ways the most impressive of the prestige neighborhoods lining US 278 west of the island.

Visionary John Reed and his merry band of developers have engineered yet another Field of Dreams on this 860-acre parcel, much as they have in their previous projects. Just like in the movie of the same name, Reed and company espouse a philosophy of "If you build it, they will come." And they do. First to Colleton River more than a decade ago, then westward to Belfair some five years later, and now to Berkeley Hall, where two Tom Fazio courses are among the centerpiece attractions.

Under normal circumstances, a pair of Fazio aces would be the main draw in and of itself. But even though these are core golf courses, with no housing or roads located within the interior of the playing area, some prospects are attracted to Berkeley for other reasons entirely. Not the least of which would be the futuristic learning center.

It's easy to picture George Jetson taking lessons in the aptly named Super Bay, easily the highlight of this avant-garde learning academy. This covered, climate-controlled hitting station features four separate video cameras documenting the swing from angles formerly unconsidered, revealing flaws in real time, slow-motion or stop-action photography that might be better left to the imagination. Balls are launched from within the structure itself, and end up on the grassland beyond the bay door. Thirty acres of pristine turf in all directions are earmarked for pitching, chipping, wedge work, sand play, and any other golf shot imaginable. Inveterate ball beaters will feel they've died, and gone to the Great Driving Range in the Sky.

Playing 7,100 yards from the tips, 6,700 from the blue markers and 6,300 from the whites, the course offers a consistent challenge, but won't overwhelm less skillful players. Literally hundreds of potential home sites were sacrificed for the core golf concept, and it's easy to envision the end result when a designer of Fazio's caliber is given free reign to create the best course he can, unencumbered by housing, roadways or infrastructure constraints. The course is top-notch from both a scenic and strategic standpoint.

Inveterate ball beaters will feel they've died, and gone to the Great Driving Range in the Sky.

Because the architect moved more than 1.3 million cubic yards of earth, Berkeley Hall's North Course offers that rarest of Lowcountry golf sensations; notable elevation change. The terrain has a rolling quality similar to the Sandhill region of North Carolina, and the serpentine course routing allows for a variety of shots both uphill and down. Nowhere is this altitude change more noticeable than at the par 3 8th hole, where the tee box sits an astonishing 53 feet above sea level, the highest point in Beaufort County.

It's a testament to the designer's acumen that the hole doesn't appear contrived, and

fits seamlessly in the midst of what might be the best stretch of holes on the course. The sixth is the first one-shot hole that players encounter. It's 200 yards slightly downhill to what appears to be a narrow opening between flowering native grasses, but ultimately reveals a generous bailout area right of the green. The next is a classic risk/reward par 5 of 560 yards, a sharp dogleg to the left. Two mediocre shots will allow a lofted approach over water to a tricky green, while big hitters can easily reach after a standout drive. After the apex at the eighth, the front concludes with a tough par 4 heading towards the clubhouse. 440 yards from the tips and flanked by sand right and water left, many players will be thankful to wobble off to the halfway house for refreshments.

The inward nine is equally stirring, and offers the same undulating greens, abundant love grasses and exacting approach shots that make the front side such a challenge. This initial course offering at Berkeley Hall fits hand-in-glove with its nearby neighbors at Belfair and Colleton River. It's yet another example of distinction in a seemingly perpetual lineup of premium golf experiences. You can keep "Beverly Hills—90210." For my money, and for serious golfers everywhere, give me "Bluffton—29910."

ESSAY: CART PATH BOUNCES

Of all the various calamities and frustrations found on the golf course, to my mind one thing stands clearly above all the rest. It's the insidious and unavoidable cart path bounce, the game's depraved way of adding insult to injury.

We are a nation of faders. All right, slicers. It's bad enough that our inability to generate an inside-out swing path in combination with our perpetually weak grips results in that pathetic glancing blow, devoid of power and chronically heading high and right. Our tee balls often cover 230 yards of airborne distance, but usually land behind the pitch mark, barely 210 yards from the launching pad. Isn't it bad enough we're always the first to hit our approaches, wearing out the grips on the fairway woods and long irons which represent our only realistic chance of reaching that par 4 green in regulation? Must we also suffer the added indignity of the ball caroming off of the cart path, heading hard right into the brush, the forest, or somebody's backyard as well?

Everyone knows that sick feeling. You come off of a tee shot and the ball balloons, then starts drifting right. You make a series of rapid calculations as time stands still, attempting to vector the angle of descent, the position of the asphalt and the wind currents as the ball hurtles earthward. A split second before impact you're resigned to your miserable fate and the ugly smack to follow, which scars both Strata and psyche concurrently.

A former neighbor and occasional golf companion once remarked that cart paths, painted green to match the grass, should be located right down the middle of the fairway. I dismissed his remark out-of-hand when he made it, but have come to realize there's a bit of method to his madness after all.

Everyone knows that sick feeling. You come off of a tee shot and the ball balloons, then starts drifting right.

First, speed of play would improve on those all too common "cart path only" days. By allowing hookers and slicers an equidistant walk to their shots, there would be far less trudging across wide expanses of fairway while carrying an assortment of clubs. Secondly, there would be the added benefit of increased distance for those accurate enough to hit the center stripe. Granted, we've all uncorked the occasional 300-yard drive when our mis-hit somehow manages to stay on the macadam, bouncing and rolling towards the green. Unfortunately, this sporadic benefit is dwarfed by the far more frequent occasion when the ball careens sideways off of the asphalt and into the underbrush.

Finally, for those who would bellyache about the aesthetics of such an innovation, I ask you this: Would a camouflaged cart path running down the middle be any more disconcerting on the course than the proliferation of condos, street crossings, power lines and roadways that already detract from what should be a bucolic walk in the park?

Of course the best solution is to do away with paths entirely. Both carts and cart paths have served to compromise, some might go so far as to say ruin the game in the last 30 or so years. It's unrealistic to expect a significant percentage of our 26 million American golfers to abandon their buggies anytime soon, though. With courses shoehorned into housing developments that require long hauls between greens and tees, the steady revenue stream afforded by cart rental, and a lazy republic tethered to their cell phones, cigars and soft drinks, carts are here to stay, and so are the paths we drive them on.

Three-putting is discouraging, leaving a sand shot in the bunker humiliating, and finding a ball resting two inches out of bounds maddening. These examples and dozens more are enough to make you turn to needlepoint as a hobby, but nothing is quite the scourge like a solid cart path bounce. I understand there are vagaries in the game, it's usually referred to as "the rub of the green." But why must we deal with the rub of the concrete as well?

COURSE DISCOURSE:
BERKELEY HALL – SOUTH

It's a striking golf course in its own right, but the most notable aspect of Berkeley Hall's South Course upon initial inspection is how different it looks from the original North Course.

Berkeley Hall sales executive Duke Delcher is a former Walker Cup player and South Carolina Player of the Year. He illustrates the differences between the two Tom Fazio "core golf" designs. "On the North Course (7,117 yards, 73.9/134) there's lots of separation because of the flowering coastal grass plantings and elevation changes. You play a hole, then turn the corner and see another, which is great in its own way. The topography of the South Course (7,126 yards, 74.5/139) is different though. There are magnificent expanses of views, under and through the mature pines, where multiple holes can be viewed at once. This is truly the parkland golf experi-

ence that was the standard throughout much of the last century."

Delcher's assessment is accurate. While Fazio's initial effort at Berkeley Hall makes abundant use of a colorfully flowering plant known as Mully grass to delineate fairways, the more recent incarnation instead depends heavily on tall Carolina pines. The original course has the most substantial elevation changes in the Lowcountry, and while the parkland style South Course features some of the gently rolling terrain seen on the earlier design, here there's only a subtle hint of changing topography. While the two courses have distinct design features, the most important element is one they have in common. They are both pure golf experiences, with no road crossings, infrastructure or housing to be built on the interior of the playing area. There's another common denominator as well. The South Course, like its older sibling, is an excellent and challenging test of the game.

It's a truism throughout golf, but choosing the proper tee box to suit one's ability is imperative at Berkeley Hall's South Course. "I'd hate to see a 20 handicap who doesn't hit it very far attempt to play the black tees," says head professional Mike Carricato. The middle markers, a shade less than 6,700 yards with a 131 slope, afford plenty of golf course for the vast majority. These blue tees are not only 500 yards longer than the whites, but on certain holes the tee box angles towards the hazards. This makes a benign hole from the forward markers a much sterner challenge, as forced carries over water or to clear bunkers enter the equation.

The par 3 holes are among the highlights here. The 183-yard third hole requires a middle or longer iron to a contoured green flanked by a large bunker, while the 150-yard fifth necessitates a tee shot with water front and left of the target. A slightly raised green 210 yards from the tee box means a wood for most players on the staunch eleventh. Only a precise short iron to the 140-yard 16th will yield a birdie putt, playing downhill to a green ringed by bunkers.

Most of the two shot holes are strong, the 414-yard seventh in particular. A bold drive over fairway bunkers will leave a considerable approach shot to a massive double green. Sharing the same putting surface as the par 5 9th, the contours and undulations here are as tricky as any on a course full of imaginative, amoeba-shaped greens. I'm not overly fond of the awkward 14th, playing just over 300 yards from the middle markers, though. Bashers

can drive the green with relative ease, assuming a straight tee shot. But those whose outer limits are in the 230 to 240-yard range are forced to aim well left of the green to find a surprisingly wide fairway that is virtually invisible from the tee. Fortunately the holes surrounding this enigma are stirring. Number 13 plays 390 yards, curving gently around a lagoon. The 15th is 435 straightaway yards with bunkers flanking water left and yet another dangerously sloping green. The occasional false fronts and ball collecting swales add more interest to these pristine Tiff Eagle Bermuda greens, a completely different surface than the Crenshaw Bentgrass greens found on the North Course. It should also be noted that the close proximity of greens to tees make this course a walker's delight, and something of an anomaly in the Lowcountry.

A brief word about bathrooms in closing. Never in hundreds of course reviews, travel features or resort profiles have I felt compelled to make reference to the on-course facilities at any golf establishment. Then again, I've never encountered anything remotely approaching the structure near the 5th tee at Berkeley's South Course. The soft lighting, classic golf photos framed on the wall and rolled cloth towels make this pit stop a comfort station in every sense of the word. It's faint praise anointing it as the finest green-grass lavatory I've seen. But what's equally true is that it has as much or more ambience than many a 19th hole or grillroom I've loitered in. It's a small detail, but another telling example of why Berkeley Hall has quickly established itself as one of the premiere destinations in the area.

SECTION IV:
BEAUFORT AREA

COURSE DISCOURSE: OLDFIELD

Number 12. Courtesy of Oldfield

It's a bit of an overstatement to say that any modern designer worth his salt has a presence on Hilton Head or in the surrounding Carolina Lowcountry. That said, the list of notable architects who haven't carved fairways through the wetlands, marsh grasses and stately oaks endemic to the area is rather short.

Not only do "The Big 3" of Nicklaus,

Palmer and Player each have several courses in the area, so do "The Big 3" of contemporary architecture; Rees Jones, Tom Fazio and Pete Dye. The area is home to courses devised by lesser players-turned-designers like Fuzzy Zoeller, Bruce Devlin, Mark McCumber and Davis Love III. There are also representations by the two most acclaimed architectural teams of the last decade, Tom Weiskopf and

Jay Morrish, as well as Ben Crenshaw and Bill Coore. Add to that other well-known names like Bob Cupp, Arthur Hills, Robert Trent Jones and George Cobb, and it's easy to see why the area is considered one of the preeminent travel destinations in the golf world.

To this glittering pantheon comes yet another stellar name. Greg Norman has unveiled his first area design, the refreshingly understated Oldfield, located on a serene and windswept piece of property midway between Hilton Head and Beaufort. "I cherish the opportunity to design a course in a home of golf like Hilton Head," states the Shark. "I like being the new kid on the block." The "new kid," as he refers to himself, has made a memorable debut.

The course at Oldfield stretches more than 7,100 yards from the championship tees and almost 6,800 yards from the back markers, with slope ratings of 137 and 133, respectively. While the humid air, generally lush turf and omnipresent wetlands around Hilton Head are perfectly suited for the aerial assault technique which is the essence of modern golf strategy, Norman has an unusual sensibility. "The ball is round, I like to see it roll," claims the architect. Oldfield is designed with the ground game in mind, similar to the way the game is played in the U.K., or in this case, Norman's native Australia. Firm, fast fairways can make the course play a bit shorter than its stated yardage, but wayward drives, and approaches in particular, will often find the omnipresent bunkering that frames so many of the fairways and fortifies almost every green. Add to that the handful of waste bunkers that Norman has implemented, and accuracy becomes even more of an imperative.

The greenside bunkering in particular is worth further discussion, as there's nothing remotely like it in the area. The architect favors stacked sod-walled bunkers with low profile entry points. "I've always been a believer that the farther a ball can roll into trouble, the harder it is to get out of trouble," states the Shark. When a player is faced with a 20-yard explosion shot over a four foot vertical wall of sod to reach the green, they'll know exactly what he means. Oldfield is a tough go for the sand-challenged. If a 20-yard explosion shot hits the wall instead of clearing onto the green, players are faced with the dicey prospect of getting up and over the edge from extremely close range, a shot that's rarely seen in these parts.

> *"I cherish the opportunity to design a course in a home of golf like Hilton Head," states the Shark.*

Furthermore, the wonderfully smooth and mostly flat putting surfaces aren't particularly friendly to those citizens of Pinnacle Nation or the Republic of Top-Flite. They'll soften some over time, but the best advice is to pony up for some Pro-Vs, Stratas or similar, softer cover ball. Depending on pin placement, there are occasional opportunities to run the ball onto the green. But a high trajectory and quick stopping power is the general recipe needed to hold greens that will repel offline or lower-arcing shots into smoothly mown chipping areas and swales around the greens.

For the most part, the designer left well enough alone on this low profile site, where very little dirt was moved in construction. "I hate to see all the artificial stuff," is how Norman puts it, and Oldfield is thankfully free of fountains, containment mounding, berms and the forced features that so many modern designers feel compelled to add to their work. One visual feature that is impossible to miss is the fabulous hardwoods throughout the property. "You've got oak trees here with 30,000 square foot canopies. When you have specimen trees like that, you have to use them, and tie them into the design." Norman is correct, as the trees at Oldfield are among the most majestic in the Carolina Lowcountry. That's no small endorsement. It's a recommendation on the order of suggesting the finest restaurant in Paris or the gaudiest hotel on the Vegas strip, and shouldn't be taken lightly.

Greg Norman captured the fifth of his 18 PGA Tour titles at Harbour Town back in 1988. Now some 15 years later he's returned once again, and engineered yet another triumph in the Lowcountry.

PERSONALITY:
ERNIE RANSOME – KING OF CLUBS

His name might be unfamiliar to most casual golfers, but Ernest Ransome III is a man well-known and much respected in certain circles. The Spring Island resident has devoted much of his adult life to the betterment of the game. The golf cognoscenti recognize his many contributions, and the lasting impact of his legacy at Pine Valley, the world's finest golf course.

I spent time with Mr. Ransome at the understated Chechessee Creek Club adjacent to Callawassie Island recently. We toured the Ben Crenshaw-Bill Coore design in mediocre fashion, but couldn't blame the weather. Blue skies and warming temperatures supplanted what seemed like a two month span of harsh conditions. The golf display was unremarkable, but stories of Ransome's 50 years at Pine Valley were anything but.

"When I joined Pine Valley in 1951 there was no initiation fee, and my dues were $50 a year. Later on it doubled though, and I was paying $100." So begins Ransome, who eventually served as Club President from 1977 until 1988, and just stepped down as Chairman of the Board a few years ago.

Ransome succeeded the legendary John Arthur Brown as President, the autocrat who ruled the club for half a century. Brown ran Pine Valley in much the same fashion Clifford Roberts ran Augusta National; as a committee of one. Members were fearful of running afoul of men like Roberts and Brown. When Ransome, at the time a young executive in his family's construction equipment business, was called to a meeting with Brown after several minor transgressions, he feared for the worst.

"In those days you could send a foursome of guests to the golf course unaccompanied by a member," recalls the New Jersey native. "I had arranged for some friends to play, and they played too slowly and held up the golf course somewhat, including Mr. Brown himself. There was a similar incident sometime later as well, and when I was called to meet Mr. Brown my assumption was I had run out of chances, and would be asked to leave the club." Little did he realize that he was being asked to join the Board of Directors, apparently to bring in some young blood and fresh perspectives for Brown and his associates, who were about 40 years his senior.

Mr. Ransome was an accomplished athlete in his youth, and played both varsity football and lacrosse at Princeton. His natural ability was evident on the golf course as well. In his prime his handicap was normally between two and five, he won the Pine Valley club championship at age 54, and once shot 68 on the course generally regarded as one of the toughest and most psychologically intimidating in the world.

Perhaps his most lasting legacy is the Pine Valley Short Course, which he designed with noted architect and fellow club member Tom Fazio. This 10-hole gauntlet is no toothless pitch and putt, and some argue it's even more intimidating than the original 18. The Short Course replicates some of the approach shots that are seen on the big course, and it was Ransome's idea to devise a way that members could practice the shots they needed in a challenging and competitive manner. Built quickly, efficiently and well within budget, the Short Course has proven to be extremely popular with members and their guests since it was introduced in 1992.

Although he'll always be most closely associated with Pine Valley, Ransome has left his mark throughout golf. He was a 1995 recipient of the USGA's Ike Grainger Award, in recognition of his 25+ years of service to the USGA. Even more impressively, a group of friends conceived the Ernest L. Ransome III Scholarship Trust at St. Andrews University in Scotland in 1994, to honor his many contributions to the game. The scholarship's mission is to provide an important educational opportunity for gifted students in the birthplace of golf, and the endowment has since grown to more than two million dollars.

Ransome and his wife Myradean have been coming to the Lowcountry for 15 years, and bought their home at Spring Island about five years ago. He's been affiliated with many of the area's finest courses during that time, including Long Cove, Belfair, Colleton River, Secession and Chechessee Creek Club. Although he never liked the idea of owning two homes, his unique status at Pine Valley precipitated the purchase of his Lowcountry home.

Perhaps his most lasting legacy is the Pine Valley Short Course, which he designed with noted architect and fellow club member Tom Fazio.

The Ransomes live in one of the few homes that are actually built within the club grounds at Pine Valley. "I had a health scare a few years ago, and realized I wasn't immortal," explains Ransome with a laugh. "Our New Jersey home is rather isolated, and we decided we should look for another home together, and not leave things to the end. We love it here on Spring Island, my wife especially. It seems we come here earlier and leave later every year. I practically have to drag Myradean back to Pine Valley."

Much as he's grown to love the laid back life in coastal Carolina, Ransome looks forward every spring to returning to his home overlooking Pine Valley's seventh hole, and its infamous bunker known as "Hell's Half-Acre." "I don't believe in ranking golf courses," he concludes, belying his former status as a Golf Digest course rater. "Pine Valley has been ranked #1 since 1985, but it's more than just the course. It's a combination of the ambience, the people, the caddies, the help and the dining room that make it such a special experience. If I knew I was going to die tomorrow though, I'd choose to play my last round at Cypress Point. It's not the greatest golf course in the world, but it's a wonderful place to sit out on the patio after the round, have a drink and listen to the seals honking in the Pacific." For the good of the game and those who love it, let's hope Ernie Ransome isn't sitting on the Cypress Point patio anytime soon.

Essay: Peaks and Valleys at Pine Valley

It was the best of times, it was the worst of times, it was the goofiest of times. I speak of my experiences at exquisite Pine Valley, the golfing Shangri-La in southern New Jersey considered by most to be the world's finest golf course.

To appreciate the absurdity of this tale, one must first understand the normal state of my game. It's best described as occasional flashes of extreme competence interrupted by long stretches of sustained mediocrity. My golf resume is as sparse as a rye grass fairway in mid-winter, and contains but a single hole-in-one, a single round at even par, and a single digit handicap ever-threatening to move to ten and points northward.

It was a wholly unexpected and delightful shock then, not only to be invited to play a golf course I never thought I'd see in person, but to overcome four sixes, several doubles, no birdies and a couple of three putts during my inaugural round several years ago and post a score of 78.

My spiked feet were ten feet off the ground as we headed to the dining room for lunch as I contemplated the magnitude of the achievement. Common wisdom at Pine Valley states that no first-timer will tour the treacherously dazzling layout in less than 80 strokes, and somehow I defied the odds and my own inabilities to do just that. It remains my single proudest moment on the golf course, and I have the framed scorecard hanging in my office to commemorate it.

Fast forward two years later. I'm making an encore appearance, this time as the guest of none other than the club's former president and chairman of the board, a member with more than fifty years of tenure. This round is the antithesis of the first, forgettable from a playing standpoint in every way, other than the day's last shot. Determined to overcome the weak fade that's left me short and right of virtually every green on the property, I overcompensated on the final approach, unleashing a vicious pull that veered thirty or more yards left of the target line. It went through the trees, into the tiny parking lot adjacent to the clubhouse, coming to rest only after detonating the rear window of a late model Mercedes.

> My golf resume is as sparse as a rye grass fairway in mid-winter, and contains but a single hole-in-one, a single round at even par, and a single digit handicap ever-threatening to move to ten and points northward.

Don't expect me to hit a medium-sized green with a pitching wedge, but if you need someone wielding a fairway wood to smash a piece of tempered glass the size of a large breadbox from 200 yards away then apparently I'm your man.

The shock and embarrassment were mitigated a bit by the odd physics of the incident. My wayward rocket somehow eluded the much taller SUV parked two feet away from the low-slung coupe, and much like the magic bullet that killed JFK, mysteriously swooped down to the target in a flurry of flying glass.

The incredible irony of the whole thing was

the identity of my victim. I was expecting to be castigated by a Pine Valley member; a captain of industry, international jet-setter, former amateur champion or some combination of the three. Instead the car's owner turned out to be another guest at the club, the executive editor of a golf magazine that I contribute to with some regularity. To his credit I never saw the slightest hint of a scowl, sigh, shrug or head shake when he learned of my misdeed. Although when I introduced myself as someone who works for his magazine, he cheerfully replied "not anymore," while offering a hearty handshake.

Only time and a timely insurance check will determine if that particular freelance outlet has dried up, but we exchanged information, I apologized profusely to all concerned, and we parted ways.

A few days later I was the guest of the professional at another superb club in the area, Metedeconk National in Jackson, New Jersey. I had apparently purged the troubling incident from my mind, the golf gods were smiling once again, and just like the initial foray at Pine Valley several years earlier, I unexpectedly walked off the 18th green having taken less than 80 blows.

My Pine Valley host is a member here as well, and I would relish the chance to someday return to Metedeconk. I'd no doubt play with unbridled confidence given an encore presentation, and it would have little to do with how well I played or the numbers adding up on the scorecard. You see at Metedeconk, there's a shuttle waiting by the final green. The parking lot, chock full of shiny, late model imports, is way over on the other side of the clubhouse, more than 500 yards away.

COURSE DISCOURSE:
CHECHESSEE CREEK CLUB

Number 5. Photo by Mark Brown

One of the problems with many modern golf course designers is their products are like McDonald's: you can find one on every street corner. Hilton Head, golf Mecca that it is, affords a perfect example. Stand in the middle of US 278 and launch a titanium driver in any direction. Your ball will likely come to rest on a patch of grass conceived by one of the modern "Big Four." It might be a Jack Nicklaus (Golden Bear, Indigo Run, Colleton River, Melrose), or perhaps a Rees Jones (CC of Hilton Head, Haig Point, Oyster Reef, Bear Creek). Perhaps it will land on a Pete Dye project (Harbour Town, Long Cove, Colleton), or a Tom Fazio (Belfair East, Belfair West, Moss Creek, Berkeley Hall). Then again, it might be one of seven local designs by George Cobb, but you get the idea.

Ben Crenshaw and partner Bill Coore are different. The two-time Masters Champion and Hall-of-Famer works slowly, selectively and chooses his projects carefully. His first Lowcountry design, Chechessee Creek Club near Callawassie Island is proof of that.

"Many modern architects draw plans, and then sub-contract out the work for the actual building of the golf course," explains Chechessee's Head Professional Franklin Newell. "Ben and Bill don't do that. They put their trust in their own shapers who've been with them for years. The men who actually shape the land know exactly what the architects are looking for, and everything is kept in house. Consequently they turn down a tremendous amount of work, because they're only comfortable doing two projects at a time. It's a totally different situation than many of the top names in the business, who might have dozens of projects going at once."

One of the many comments that Newell gets from first time players at Chechessee is that the course looks like it's been there for 75 years. This course, delightful and challenging as it is, is no modern dazzler. The maturity of the trees, the lack of artifice, dearth of modern mounding and the organic manner in which the holes fit the landscape make Chechessee one of the most natural looking tracks in the Lowcountry. The course scores further points because it's completely walker-friendly with a vibrant caddy program, and volume of play is minimal. Currently, the private club averages about a dozen groups per day, and even when fully subscribed will probably have an annual round total of 12,000 rounds or less. This is throwback golf as it's played none too often in these parts, with everything you need and nothing you don't.

Befitting a course designer who's considered the finest putter of his generation and one of the greatest putters in history, much of Chechessee's bite is around the greens. A host of pushed-up greens, many with straight edges and almost all with severe slopes and undulation, are what help defend par at a course which is barely 6,600 yards from the back tees. Crenshaw and Coore were influenced by the work of Seth Raynor, the early century architectural master whose nearby work includes Country Club of Charleston and Yeaman's Hall, also in the Charleston area. A former landscape architect who preferred to work in straight lines, Raynor was known for his distinctive straight-edged greens, and Crenshaw-Coore have followed suit here in the Lowcountry.

Many players will be thankful that the golf ball will be easy to drive and hard to lose, as fairways are quite wide and hazards are at a minimum. There's great variety in the holes on the course, with a trio of par 4s that are under 350 yards, and four more that are better than 435 yards. Five one-shot holes, varying in length from 165 to almost 245 yards from the back markers, offset three par 5s, with a resulting par of 70.

Until Chechessee Creek Club debuted, the nearest Crenshaw-Coore design was Cuscowilla, about three hours east of the area at Lake Oconee, Georgia. I've never played Sand Hills, Crenshaw's Nebraska masterpiece that occupies the #1 ranking in Golfweek's list of Modern American courses. I have been fortunate to play Cuscowilla several times though, which is ranked a few notches below Sand Hills, hovering just outside the nation's top ten.

Chechessee Creek lacks the elevation changes of central Georgia and the dramatic reddish bunkers that hallmark the property, but little else. It's not surprising that Chechessee Creek has already earned a place on Golfweek's "America's Best" list of modern courses. It's that good.

Essay: Planes, Trains and Automobiles

One of the first newspaper columns I ever wrote was actually a contest. In attempting to engage the subscribers, I offered a free golf foursome to anyone who could identify the biggest golf nut in the Lowcountry, providing they proved it in an essay recounting their nominee's exploits. That inaugural contest winner described a true fanatic, but I realized later that sometimes you find what you're looking for right under your nose.

The fact is that one of my very good golf buddies is crazier than anyone else I've met in the game, and would be the hands-down winner here in the Lowcountry, the high country, or virtually any country. Let me tell you a bit about Captain Rivi, and why if I hadn't been in charge of the golf nut contest myself, my candidate would've won it in a landslide.

The Captain is eccentric in several regards, not just his golf exploits. For example, he wanted to see Maple Leaf Gardens in Toronto before they tore it down a few years back. So he drove alone from Philadelphia, 15 hours round trip, to see a game. Twice. He took the train to Baltimore, sans ticket, to see Cal Ripken break Lou Gehrig's consecutive game streak. He bought a ticket on the street, amazingly enough at face value. He snuck in without having it torn, and now has an invaluable keepsake in addition to the memory. He's the kind of guy who'll get up at any hour and drive any distance to wait by the starter's shack in case there's a cancellation at a golf course he heard was worthwhile. When I invited him to take a tour of some of the finest private clubs in our home state a few seasons ago, he was at his very best. Here's a slice of life that I witnessed personally, as the irrepressible Captain Rivi did his best to balance golf, work and the home life in one incredible four-day stretch.

On a Sunday, he got up in Philadelphia, and drove almost five hours to join me for a round in western Massachusetts. Neither of us could come within hailing distance of the course record there, still held by Bobby Jones, but we celebrated with a beer nonetheless. He then drove three hours back to New Jersey.

Monday morning found him briefly in Manhattan, ostensibly to "show his face" at the office. He was back in western Massachusetts by noon though, to pick me up and head towards Boston. It should be noted that the Captain's decrepit auto, which thankfully has since been replaced, demeans the word "jalopy." Among other indignities it was devoid of air conditioning. Nevertheless he arrived in 90 degree heat in his suit, tie firmly knotted, after another three-hour drive from the big city. We proceeded eastward on the turnpike, scalding air rushing in the windows. We were given a quick tour of The Country Club in Brookline, which at that time was in final preparations for the Ryder Cup Matches. We then proceeded to play another old-line Boston club in the late afternoon.

The Captain absolutely had to be at work on Tuesday. He planned to drop me off with family well north of the city, then drive to New York late that night. Instead he decided he needed to get home to see his wife, so instead he rushed to Logan Airport to catch the last flight home to Philadelphia instead. Let the record state that he planned to drive to Philly had he missed that last flight, and there's no reason on earth to doubt him.

He took the two-hour train ride to New York the next morning, his normal commute. He worked all day, and then boarded a train bound for Boston in late afternoon, a train which was delayed interminably. He took a cab to Logan and found his car. He drove 45 minutes to where I was staying and arrived, as God is my witness, at midnight, suit still on, tie firmly in place.

The fact is that one of my very good golf buddies is crazier than anyone else I've met in the game, and would be the hands down winner here in the Lowcountry, the high country, or virtually any country.

We played golf once again on Wednesday morning, accompanied by caddies on one of the finest courses in New England, a very recent host of the U.S. Senior Open. Our round finished just past noon, and though we had missed breakfast, the Captain declined my offer for a quick lunch. He made an immediate departure for New York and the office. Presumably he got his suit on, tie firmly knotted, between tollbooths on Interstate 95.

The Captain's golf game has improved dramatically in recent years. The banana ball has straightened considerably, the predilection for missing two footers all but disappeared. He's still got driver-phobia though, and prefers to hit the three wood on more holes than necessary. What's odd is that even though he doesn't get the distance he needs on the tee shot, his drives are exponentially longer than anyone else I've ever golfed with.

Essay: The Secession Story

Savannah residents Bill Degenhart, Steve Yeager, Jules Victor, Sidney Bolch III and Michael Hemphill have several things in common. They are all doctors, native Southerners, and serious golfers with single digit handicaps. They are also members of the same golf club, a little-known gem that lies just an hour north of the Savannah River Bridge, in bucolic Beaufort, South Carolina. Their club is unfamiliar to the vast majority of golfers in the world. It has had a short history, has never hosted a tournament of real consequence, or been featured on television. But in certain pro shops and grill rooms across the land, where the golf cognoscenti gather at the most venerable golf institutions, a single word identifies this club. Discriminating players nod in recognition and appreciation, and smile. The word is Secession.

The Secession Club differs from most other golf clubs because of three things it *doesn't* have. It doesn't have golf carts, it has relatively few members of average golf ability, and it has a paucity of local members. Mike Harmon, a former PGA Tour pro, is the Director of Golf at Secession. He reflects on the hard-core constituency that makes up the club's membership. "Our members are unique," begins Harmon. "They join here knowing that there are absolutely no carts allowed. If they blow out a knee or their back, then they must pass their membership to a son, or sell it back to the club. There is no other recourse. We have two golf carts here, which are used on a temporary basis by members who have legitimate medical excuses. Part of my job is to make sure that number never goes to three."

At Secession, if you want to play golf, then you must walk. And you must walk with a caddie. This 750-member private club boasts not only a national membership, but also a national caddie corps. The caddie stable includes aspiring or former playing professionals as well as veteran loopers from some of the most fabled tracks in the country. Places like Medinah in Chicago, Winged Foot near New York City, and Pine Valley in southern New Jersey are proving grounds for many a Secession caddie.

Secession is one of a small handful of truly national clubs. About 65% of the membership resides in the North, and the remaining 35% come from cities in the South.

The walking-only, caddie-only policy attracts a different breed of golfer than you'll find at other clubs. Consider that the average handicap for a golfer in the United States is 18. By comparison, about two out of every three Secession members has a single digit handicap. About 150 of them play to a 3 or better. Savannah native Jules "Bubba" Victor, one of only 50 local members, explains the appeal of the club. "Secession is not a typical golf course, nor is it a typical golf club," begins the Chief of Staff at St. Joseph's and Candler Medical Center in Savannah. "It is golf in its purest form. There are no swimming pools, tennis courts, or real estate developments near the course. The reason there are so many skilled players there is because good golfers beget good golfers, in my opinion. The club is invitation only, so it's understandable that strong players invite their friends to join. It's not to say that there are only skilled players at Secession, but everyone there has a real love of the game and its traditions. People are attracted to the club not just for the golf, but for the purity and sanctity of the game itself."

Mike Harmon came on board in 1987, a full five years before the golf course was completed. He agrees with Dr. Victor's assessment of the unique qualities of a Secession member. "The idea of a tradition-laden club held tremendous appeal. We've never advertised for members, because our policy has always been invitation only. We pre-sold 300 memberships before construction of the golf course itself even commenced. If we had wanted to compromise our philosophy and advertise the club, we probably could have sold out in a year, but the fact is we've added members slowly. If we had advertised, we would've gotten people who would like to *say* they are Secession members, as opposed to actually *being* members of the club."

Secession is one of a small handful of truly national clubs. About 65% of the membership resides in the North, and the remaining 35% come from cities in the South. The local members come predominantly from Hilton Head and Savannah, with a smaller representation from Charleston and Beaufort. "We decided initially that we would only have 50 local members (residing within 100 miles) because we wanted to make this a true national club," continues Harmon. "Our local members are special to us, because they are all traditionalists, and they want an affiliation

with a club like ours. A guy from New York joins because of our good weather and the club's unique qualities. Our local members don't need to join because of the weather, but because our philosophy about the game meshes with their own. We chose to locate the club in Beaufort because it is a good distance from the resort atmosphere of Hilton Head, but at the same time is easily accessible to the Savannah airport. This is a key component, because the vast majority of our members fly in for a few days at a time."

The Secession club was initially cast as a Pete Dye course, but Australian pro Bruce Devlin delivered the final product. The Director of Golf explains the circumstances which led to Dye's replacement on the project. "Pete Dye and his son P.B. were hired at the beginning, back in 1986, to design this course. By the time we were getting the construction underway, Pete was committed to the Ocean Course on Kiawah Island. This was in '88 and '89, the Ryder Cup was coming to Kiawah, and Pete made that project his priority. P.B. took over the day-to-day operations, and his ideas clashed with the original partners over many of the design concepts. The Secession founders wanted a subtle, natural-looking, links-style course, with little earth moved. They didn't feel that P.B. was moving the project in that direction, so Devlin was brought in to complete the construction. He was an excellent choice, because he has plenty of experience, and has a real knowledge of links courses because of all the time he's spent overseas." The partners got what they wanted. The course was constructed with minimal earth moving; only about one third as much as typically seen in modern construction.

"Bubba" Victor maintains a hectic schedule, but relishes the two or three rounds he typically plays at Secession in a given month. "That club is like many special things in life, in that you don't want to overuse it. It's a golf course you could play every day, because it's always changing. Unfortunately, I don't get up there as often as I'd like, because I'm so busy. I don't think I've ever been up there, whether I've played good or bad, that I've not had an exceptional experience." Victor picked an opportune time to be playing well several springs ago, when he captured the Secession Club championship. "I'd have to put that club championship at the top of my list of golf accomplishments," concludes the former captain of the University of Virginia golf team. "To have my name on the Champion's Board at a club of this caliber really sends a chill down my spine."

Harmon is pleased with the way Secession Club has evolved in its short history. "This is a true golf club, built out of the love of the game, and modeled after some of the great old clubs around the country. There has been a wonderful resurgence in recent years of new clubs like ours, which attempt to incorporate the best elements of our finest courses; places like Augusta National, Cypress Point and Pine Valley. Secession is a wonderful place to walk and play golf. How great a golf course this is isn't for me to say. But I will say it is a wonderful place to walk and play."

COURSE DISCOURSE: SECESSION CLUB

Number 16. Photo by Mark Brown

Secession is unlike any other golf course in the Lowcountry. Places like Harbour Town and Melrose leave a vivid first impression, but not Secession Club. Unless a golfer has played in the British Isles or is familiar with links-style golf, Secession will be an unusual golf experience. There are absolutely no houses, very few trees, and a landscape that lacks defining features. This is a true member's course; it is the type of layout that reveals its strength and intricacies over time and repeated playings. It is the antithesis of a resort or high profile course. One doesn't walk to the clubhouse after an initial round marveling at the tricky design features or remarking on the majestic signature holes. Secession is a golf course that is not easy, but simple and mostly straightforward. Not boring, but subtle.

Secession plays to a length of just under 6,700 yards from the middle tees, and carries a sturdy slope rating of 134. The course has two major lines of defense; the wind and the marsh. Both of these features are capricious and unpredictable. The wind can blow, whip, howl or disappear completely. It can come from any and sometimes every direction. The marsh can be just as fickle. Present on more than half the holes, a wayward shot can easily be played out of these lateral hazards. Assuming the ball can be found. Some will come to rest sitting up nicely on either hard packed sand or on top of the straw. Others will be buried, unplayable, irretrievable or lost in the muck and vegetation. It's all in the luck of the bounce, the phase of the moon and the height of the tide. It is the rare player, no matter how skilled, who can negotiate all 18 holes without at least a single visit into the marshland.

Unless a golfer has played in the British Isles or is familiar with links-style golf, Secession will be an unusual golf experience. There are absolutely no houses, very few trees, and a landscape that lacks defining features.

it also thankfully lacks many of the bells, whistles and design gimmicks one often sees in modern golf course construction. It is an exceptionally fine course, and is in many ways like a bottle of fine wine. It is a golf course one acquires a taste for, and grows to appreciate more as time goes on.

Balls that are in the fairway can be played towards the green in a number of ways. Most every green has an opening in front, so the bump and run can be as much a part of one's playing strategy as the aerial assault. The greens and aprons that front them are quite hard, so it's best to land a shot before the green and let the ball roll towards the flagstick. The putting surfaces themselves are pristine; they offer a true roll and are not overly humped or contoured. Recent modifications include a smattering of stacked sod-wall bunkers added to both the landing areas and greens. These hazards can be extremely penal, particularly if one is forced to play out laterally to extricate the golf ball. Broom-straw, a thick strain of vertical grass, has also been added to help frame certain fairways and complicate recovery from particularly wayward shots.

The Secession Club is an unusual, understated golf course. Little earth was moved during construction, and the setting is natural and tranquil. Secession is not the type of course that is necessarily going to appeal to everybody. Not only is it lacking cart paths,

PERSONALITY:
MIKE HARMON – SULTAN OF SECESSION

Beaufort's Mike Harmon is a good man with a great job. As the Director of Golf at the Secession Club, Harmon oversees the day-to-day operation of one of the most unique and desirable golf clubs in the country. I was able to join Harmon for a round at Secession recently, and learn more about the man and his background.

"I never had a stellar professional career, or amateur career, for that matter," begins the former PGA Tour player. "I was slightly in awe of the other players on Tour when I got out there, and that's no way to make a living. My first tournament as a Tour pro was the U.S. Open at Baltusrol in 1980. I practically wanted to go in the clubhouse and get autographs. A well-seasoned amateur or pro doesn't think that way."

Harmon played the Tour for three seasons, 1980-1982, and managed a few top 25 finishes during that time. The closest he came to victory was a runner-up finish in the 1980 Walt Disney World Team Championship. He and his partner finished three shots out of the lead, when a victory would have resulted in a four-year exemption on Tour. "I

missed keeping my card by three shots over six rounds during qualifying school at the end of the '82 season," continues Harmon. "A friend of mine named Brooks Simmons was the Director of Golf at Palmetto Dunes, and he offered me a job for the winter. That's how I ended up in this area."

Playing golf with Harmon at Secession is like taking on Larry Bird in the Boston Garden, or Pete Sampras at Wimbledon. The man has a serious home field advantage. His swing is smooth as syrup, but the ball comes off the clubface like it's been turbo-charged. He hits every shot with a draw, and as the ball drifts left towards the green it's usually seeking the flagstick. Thanks to his three- putt on the par 4 first hole, I took an unlikely lead, but it was as short-lived as a Chevy Chase talk show. Harmon pummeled me into submission with an unrelenting string of pars, and then changed putters at the turn so he could start banging in birdies. In golf accuracy is key, but Harmon proves that power is paramount. An objective example: Holes 11, 12 and 13 at Secession are a trio of dificult par 4s, averaging almost 435 yards each. With the wind howling on this exposed terrain in characteristic fashion, Harmon attacked the greens with an 8 iron, 5 iron, and 9 iron, respectively. I guess we could all threaten par by holding a mid or short iron on every hole. I tried to finish in high style, curling a 25-foot birdie putt around the rim on 18. The director did me one better; canning a 20-footer a moment later, his third bird on the back side. I waved the white flag, and we retired to the veranda.

> *"I take more pride and satisfaction in having helped put a place like Secession on the map, than I do from anything I ever achieved as a player."*

"I knew that if this club worked out, that the golf director's job here would be one of the best in the country," continues Harmon, who spent two years as the head pro at Moss Creek Plantation before coming to Secession in 1987. "Not just monetarily, but as far as the environment and membership were concerned. Because we're a national club, our members only come down a few times a year. They can't wait to get here, and it's like having a reunion every week. It's much different than a local golf pro that sees Mr. and Mrs. Jones walking through the pro shop four or five days a week. This job never gets stale, and it's worked out very well."

Mike and his wife Lynda, both native Atlantans, have a preteen daughter named Joanna. "This is an ideal location for us. It's close to home, and it's a beautiful place to live. A few years ago we bought a small ski house up at Beech Mountain, North Carolina. We go up three or four times a year, and ski about 15 days a winter. It's really the best of both worlds."

Harmon is in an enviable position as the only Director of Golf the Secession Club has ever known. He plans to keep that distinction for quite a while. "We took our time, and made sure we invited the right people to join the club. It's been a long and difficult road at times, but we've just recently filled our membership roster at 750. Now I'm looking forward to playing more golf, spending more time with the membership, and continuing to work on my game, which has started to come around." A recent tournament victory in a significant pro-member event at nearby Haig Point, his first in a decade, and a flirtation with the Secession course record prove that Harmon is once again finding the form that put him on Tour in the first place. "I take more pride and satisfaction in having helped put a place like Secession on the map, than I do from anything I ever achieved as a player. I'm proud I made it out on Tour, but what we've accomplished here is more lasting. It gets to the heart of what the game is really all about."

SECTION V:
SAVANNAH

PERSONALITY:
GENE SAUERS – BACK FROM THE BRINK

"Gene Sauers is a has-been," crowed the barfly in the 19th hole, tacitly admitting his own status as a never-was. Ordering yet another beer, he exclaimed, "he was steady out there for years, but now he's just gone and disappeared."

Those cutting remarks might've been accurate when uttered, but were made prior to September 1, 2002, arguably the most important date in Sauers' nineteen years as a professional. It was the day the Savannah native made a dramatic reentrance on golf's main stage, and his implausible triumph at the Air Canada Championship vaulted the Tour veteran from semi-

obscurity back to fully- exempt status, a rarified privilege he hasn't enjoyed since 1996. "It's going to be great to be back out here where my friends are," exulted jubilant Gene after this most unlikely of victories. "It's a dream come true."

Indeed, Sauers sightings on the PGA Tour had become almost as rare as double eagles over the years. The Canadian win was only his 20th Tour start since 1999. The dwindling appearances since he lost his exempt status after the '96 season; 16 starts in '97, then 11, 6, 8 and 2, provide conclusive evidence that both his game and name weren't impressive enough to command the sponsor's exemptions he had hoped for. Instead he was doing time in the bush leagues of the Buy.com Tour, more than 75 appearances in the five year stretch between '98 and 2002.

The frustrations were evident for the long-time PGA Tour vet back in the spring of '99, when he complained to Sports Illustrated that the lack of courtesy cars and free food in the clubhouse hindered performance for those mired in the purgatory of golf's secondary circuit. He was called on the carpet for his remarks, but in truth it was difficult to give up the pampered life of the top echelon pro for the scruffier environs of the minors.

Long-time area residents or maniacal golf fans might recall that Gene was a cash machine for almost a decade in the mid '80s through the early '90s. He was steady if not spectacular, and increased his earnings in each of his first nine seasons, won twice, and always finished between 31st and 42nd on the money list. He also averaged about $320,000 per year in his best seven seasons. Things slowly went south though, and it appeared a fairly safe bet that Sauers, on the wrong side of 40, would never be heard from again on golf's grandest stage.

It would've appeared that way at least, to almost everyone but his wife Tammy. "It may sound funny, but I don't feel he came out of nowhere," she explained shortly after the life-altering victory. "Gene's not a partier, he doesn't touch coffee on the road, and we've taken up yoga together to work on relaxation. I always knew he could do it."

It's a whole different world for the Sauers family now. Gene and Tammy's three sons, Gene Jr., Rhett and Dylan weren't yet born

when he last won in Hawaii in 1989, and likely have little recollection of their dad as a bona-fide Tour player back in '96.

Tammy's synopsis of the week leading up to the win in Canada is instructive, showing how the netherworld status of a non-exempt player can make the already hectic life of a touring pro exponentially more difficult.

Sauers sightings on the PGA Tour had become almost as rare as double eagles over the years.

"Gene was on the road for five straight weeks, three events on the big tour and two on the Buy.com. He got home late Sunday night past midnight, and then got the call Monday afternoon that he made the field in Vancouver. (He had begun the week as the seventh alternate.) He had to grab a 6:45 flight out Tuesday morning, but he bought a one-way, walk-up ticket to fly, because he didn't know if he would be going to the Buy.com event in Utah or the next PGA Tour stop in Toronto. Needless to say, a purchase like that is a red flag at the airports these days, so even though they know him after all the years of travel, they totally went through his luggage before the flight."

Sauers overcame a late arrival, delayed baggage and an opening-hole bogey to win for the third time, bank $630,000, more dough then he ever did in his best season, and almost as much as he's won in the last eight years combined. But the windfall was secondary to the two-year exemption, the keys to an ever more lucrative kingdom where he'll now have full access though 2004, and the chance to build mightily on career earnings that are already past the four million mark.

"I knew down deep that I still had the talent to compete and win out here," explained Sauers in the aftermath of victory. "But I wasn't positive I could reach deep enough, work hard enough to make it happen."

"The sky's the limit for us," concludes Tammy. "He may be past his fortieth birthday, but I think he's just getting started. I guarantee he's going to appreciate his Tour status even more this time around, and not take anything for granted." The win provided a turbo-charged boost to the Georgia Southern grad's world ranking, vaulting him an almost comical 740 places, from 895th up to 155th in a single week. Ever the sensible one, Sauers savored the win from a practical standpoint. "Now I'll be able to plan my schedule in advance," he concludes with a sigh of satisfaction. "Maybe get some cheaper airfares."

The Whitemarsh Island family looks ahead, not only to a settled travel schedule, not only to appearances in many of the marquee events the boys have never known their dad to be a part of previously, but also to a potentially dazzling future, a future that was beyond the outskirts of imagination just a short time ago.

ESSAY: SO YOU THINK YOU'RE GOOD? YOU'RE NOT!

So you think you can play a little, do you? We've all heard that 90% of golfers can't break 100, or 100% of golfers can't break 90, or other, similar examples of statistical drivel, so you low 80s shooters have a right to be a bit smug.

Perhaps you're a notch or two better than that, though. Maybe you score in the 70s with great regularity, keep a handicap in the low single digits and have to shuffle your member-guest invitations like a deck of cards.

Are you even better than that? Are you a former junior standout or collegiate player, a gent comfortable shooting par or close to it, in the 60s on rare occasion, accustomed to playing from the tips and reaching par 5s with irons? I've got news for you, Ace. In the overall scheme of things you're still not that good.

I draw these conclusions after a round with a young man you've never heard of and perhaps will not hear of again, although he'll make every effort to attain a degree of golf celebrity in the years to come. Larry Nuger, the son of my friend Steve Nuger, is in his early twenties and is a former scholarship player at the University of Illinois. Larry has a couple of major highlights on his resume to this point. They include winning a collegiate event by eight shots, co-medalist honors in the Big 10 Tournament, second place in the New England Amateur and a similar runner-up position in a recent Massachusetts Amateur, losing a chance for the title on the 36th and final hole.

I won't bore you with the details of young Nuger's talents. Suffice it to say the ball comes off his driver in a thunderclap, approach shots streak through the sky like bolts of lightning, bunker explosions are soft as fog and putts rain down from all angles and distances. He's got superior command of all facets of the game, if not the weather itself.

Larry has graduated from Illinois and recently joined the play-for-pay crowd. It remains to be seen if the combination of passion, power and precision he brings to the game will meld ineluctably, bringing him some measure of fortune, and perhaps one day into our living rooms as well. A world of opportunity awaits, with various professional tours available as on-the-job training. Futures and Hooters, Golden Bear and Golden State, perhaps the Nationwide Tour and if fate would have it, one day the Big Show.

A world of opportunity awaits, with various professional tours available as on-the-job training ... Or perhaps not.

Or perhaps not. Wishing him success goes beyond the bragging rights attendant with announcing "I knew him way back when," as he's truly a pleasant and personable fellow from a fine family. But facts are facts and odds are odds, and to this point Nuger is but a footnote, a blip on the radar screen in the world of low-level professional golf. He's six months younger than Matt Kuchar, six months older than Charles Howell III, and dearly hoping to someday join his contemporaries on the game's main stage.

Even if Nuger never earns a crust though, his future seems all but assured regardless.

There's more than a bit of room in the business world for a bright and industrious kid who's a helluva stick besides. You're making quite the impression when you're breaking par, especially when paired with industry titans sporting enormous influence and responsibility, with handicaps to match.

So polish that club championship trophy. Take pride in seeing your name emblazoned on the President's Cup plaque in the 19th hole. Stuff that skins money deep in your pocket after the clutch up-and-down on 18, and enjoy the aroma wafting from your victory cigar after the big Member-Guest. But remember that Larry and a faceless thousand just like him can give you four shots a side and still finish you on the fifteenth. Try and keep it all in perspective.

COURSE DISCOURSE:
THE CLUB AT SAVANNAH HARBOR

Number 17 . Courtesy of Savannah Harbor

Crossing the magnificent Savannah River bridge towards South Carolina, experienced drivers turn their gaze to the right. Looking east affords a glimpse of the leafy and flourishing downtown, while looking west provides a landscape of industrial blight. Now there's another compelling reason to fix your eye on the east. It's the Club at Savannah Harbor, a sterling Bob Cupp and Sam Snead co-design located on a spit of land on the Savannah River called Hutchinson Island.

This commendable layout is part of the Savannah Harbor Resort, and many players stay on premises at the 403-room Westin Hotel, which is one of the finest lodgings in the Lowcountry. Presumably so will gentlemen like Hale Irwin, Tom Watson, Ben Crenshaw, Tom Kite and their colleagues when they come to town. Though a fairly new venue, Savannah Harbor has made a distinctly positive impression, and has been selected to host one of the marquee events on the Champions Tour, the Liberty Mutual Legends of Golf, for at least four years, beginning in April of 2003.

You won't find too many vacationers, or Champions Tour pros, for that matter, willing to tackle this track from the laughably long championship markers, just a dozen steps shy of 7,300 yards. That's a reasonable distance for a golf course at altitude, something in Colorado or Utah. Here at sea level in the humid air, it's borderline psychotic for all but the strongest players. The middle boxes are more than enough for most decent golfers, stretching over 6,600 yards. The front tees are perfect for seniors or "resort players" (i.e., hackers) at about 6,050.

As befitting a resort course, the first few holes serve as something of a warm up. A medium-length par 4 is followed by an equally benign par 5, both with generous landing areas. The third hole is a change in tone though; a 430-yard par four with a monstrous, 17,000 square foot green. With about 65 paces from back to front, the various pin placements can result in as much as a five or six-club difference on the approach shot. Frustrating as it might be to be faced with a 50-yard putt, the sheer size of the green and consequent variety of pin placements ensure a golf hole that will dramatically play differently from day to day.

The short sixth hole is a straightforward par 3. It's unremarkable, save the unforgettable view of the Savannah River bridge looming majestically behind the flagstick. One expects the bridge to be the dominating structure on the landscape, but in fact the most noticeable structure on the premises is the gleaming hotel. The bridge is more of an innocuous presence; a player can enjoy the graceful span or choose to ignore it.

More disconcerting are the other urban encroachments that detract just a bit from an otherwise bucolic experience. There are inherent difficulties in a golf course located in such close proximity to a city center. It's nice to be a stone's throw from downtown,

but by extension, you're also shoulder-to-shoulder with the industrial base. The occasional sight of smokestacks, pulp mills and power lines tempers the novelty of playing golf within view of the bridge or the golden dome of city hall.

Though a fairly new venue, Savannah Harbor has made a distinctly positive impression, and has been selected to host one of the marquee events on the Champions Tour, the Liberty Mutual Legends of Golf, for at least four years ...

On the other hand, with the exception of the trio of courses found on Daufuskie Island and just a few other examples, finding a truly rustic golf experience in this area is virtually impossible. A never-ending succession of houses and condos line almost every fairway at the Landings Club on Skidaway Island and Harbour Town Golf Links in Sea Pines Resort, just to name a handful. In time, thoughtful tree plantings and landscaping will help to temper much of the urban intrusion which currently detracts a bit from the "walk in the park" ideal of a day on the links.

As benignly as it begins, the golf experience at Savannah Harbor concludes in stalwart fashion. The 15th and 16th are particularly harrowing. First is a long par 4 with wetlands guarding the right side as you approach the green. The next is a gargantuan par 5, 600 yards from the middle tees, with a visually intimidating drive. Many players will be in the unusual position of hitting a wood over wetlands to approach the green if their tee ball and second shot are anything less than ideal.

The shortish par 3 penultimate hole offers some relief, though it's ringed by bunkers. The final hole is a par 4 playing directly towards the clubhouse. It has marsh to the right, a copse of trees to the left and numerous bunkers guarding the green. Not exactly a stroll to the finish line.

Closing stretch aside, the Club at Savannah Harbor is a challenging, but not overwhelmingly difficult test of the game. The driving areas are generous, and while far from a treeless expanse, the tree cover is mostly incidental, much like the bridge in the backdrop. The course is managed by Troon Golf, one of the best names in the business, so conditioning is almost uniformly excellent.

Savannah's golf options have always paled in comparison to the richness and diversity of the Carolina Lowcountry. The Club at Savannah Harbor, a welcome and meticulously manicured addition, helps to tip the scales just slightly back in balance.

PERSONALITY:
LEO GRACE – STATE OF GRACE

You might not know Leo Grace by name, but he is a man who personifies all that's good about the game of golf. Leo is a moderately tall, distinguished looking gentleman with gray hair. As a young man he lost his left eye when a tumor was removed from his orbital cavity, and consequently he now wears a cotton gauze patch.

Leo is an even tempered man with a smooth, unruffled golf swing. His fluid technique stands in stark contrast with the hurried lurches so often displayed at local driving ranges. He's not prone to discussing the golf triumphs of his younger days, but the fact is he was the Massachusetts State Junior Champion some 50 years ago, and was several years later a finalist in the Massachusetts Amateur. Leo's game rebounded quickly after his surgery. The tumor was removed during December of his junior year at Boston College. In May, he won the first of two consecutive New England Inter-Collegiate Championships, and was eventually inducted into the BC Athletic Hall of Fame.

Leo and Jean Grace raised six children back up in Boston, and migrated south to the Lowcountry several years ago. Since we were introduced, I've been fortunate enough to play golf with Leo on several occasions, but one round stands out in particular.

One autumn I was entertaining a couple of visitors from New England, frequent golf companions of mine, with respectable handicaps not far from single digits. I asked Leo to join us to complete the foursome, and he delivered a five-birdie, five-bogey performance for a seamless 72. My guests were confounded by the overhanging trees which lined so many of the golf course fairways, and neither could manage to break 90. Meanwhile, quiet Leo, old enough to be anyone's father, easily maneuvered his way to an even par round. A couple of times that afternoon he dropped lengthy putts of more than 25 feet, and was shaking his head and laughing as the ball disappeared into the cup. But as we played the final few holes he wasn't nervous, or waiting for the spell to break. He was in control of his game and of his swing; it was obvious he had been at par or below countless times before, and never seemed to be out of his element.

One moment that day stands out in my memory. I was tending the flagstick, and Leo putted the ball to tap-in range. Without thinking, I reached down, picked it up, and tossed it to him from about 10 or 15 feet away. While the ball was in the air my heart started to pound, and everything seemed to go in slow motion. I was thinking, "What if he can't see it that well? He's probably not used to catching thrown objects, I don't want to bonk him on the head..." My worries were groundless. Leo snared the ball like a major league first baseman who fields a throw from across the diamond, simultaneously watching the progress of an advancing runner. He caught it one-handed, and was looking at me as the ball landed in his palm. I breathed a sigh of relief, and in retrospect considered it one of the most impressive things I had seen all day.

There are plenty of senior golfers in the area who possess at least as much skill as Leo Grace. Doubtless there are other accomplished players in the area who have overcome disabilities and physical hardships as well. I mention Leo because to me he is a picture book example of why golf, at its essence, is truly a simple endeavor. Swing smoothly, think clearly, play conservatively, and your score will add up slowly. He's a solid citizen, with a solid game to match.

ESSAY: LIFE IN THE LANDINGS

It took us about six months to settle our affairs and leave New England for coastal Georgia. During the interim between announcing our impending departure and the departure itself, I would field the question "Why Savannah?" from countless friends, business associates, neighbors and acquaintances. I would offer a boilerplate response about the region's climate, coastline, culture and character, but a reluctance to reveal too much kept me from telling all but my closest confidants the true reason. "You should see the neighborhood we're moving to."

If it weren't for The Landings, we might've eschewed Savannah for Montana, Indiana or Susquehanna. But the fact is that The Landings was the real reason we headed south; the chance to set up shop and raise the kids in one of the most beautiful, peaceful and unspoiled environments on the East Coast, if not the entire nation.

"Disneyland for Adults" is how some folks describe the community, a 4,500-acre paradise on Skidaway Island, just east of Savannah proper. The physical attributes are impressive in their own right, but life at The Landings goes far beyond six championship golf courses, dozens of tennis courts, four clubhouses, two marinas and miles of biking trails. Traffic is minimal, courtesy the rule, litter practically nonexistent. It's the type of place where you can leave an unlocked bicycle unattended for hours, and find it undisturbed upon return. Its residents are almost uniformly people of accomplishment, but egos are generally checked at the door. It's a place full of former CEO's, company presidents, high profile executives and international businessmen, but remarkably few pull

rank, and rarely will you hear someone braying "don't you know who I think I am?" to a cowed waitress or assistant golf pro. In describing human nature it's often said you can find bad apples anywhere. Not just at fancy country clubs, but at gas stations and grocery stores as well. Well, the area gas stations and grocery stores must have a disproportionate share of jerks, because at The Landings they are scarce, the vast majority of 7,000 residents altogether pleasant.

If it weren't for The Landings, we might've eschewed Savannah for Montana, Indiana or Susquehanna.

If you were painting a picture in broad brush strokes, you could refer to The Landings as a retirement community. Technically you'd be correct. But most so-called retirement communities are devoid of ballfields, playgrounds, jungle gyms and waterslides. The fact is that there are hundreds of families who make their home on Skidaway Island, and many hundreds of children; infants, adolescents and teenagers. Besides the bicycles and birthday parties, swim teams and school busses, summer sports camps and Santa visits, there is one other definitive.

Proof-positive comes every Halloween night, when a time honored ritual is enacted that has to be seen to be believed. There are no statistical tomes or reference books that can lend assistance in this manner, but I'm reasonably secure in the veracity of the following statement. The Landings is one of the very few places on planet earth where Halloween trick-or-treating is conducted via golf cart. It may not rate with the Running of the Bulls in Pamplona or Carnival down in Rio, but October 31st on Skidaway Island is a scene to experience nonetheless.

That said, the fact of the matter is that most folks in the neighborhood are old enough to be my parents, and that can get a bit stifling over time. It was culture shock early on, when I asked an occasional golfer pal of mine, a distinguished World War II veteran, what his oldest son did for a living. "He's retired," I was told. Fortunately we spend a good deal of time out in Park City, Utah, which is one of the most dynamic skiing, biking, hiking, rafting, outdoor oriented communities in the nation. My wife and I marvel at the juxtaposition. We transcend from among the youngest to among the oldest in a single plane ride.

In the course of our five-plus years in Savannah, we've been visited by dozens of friends and family members, as well as acquaintances who've been traveling though. We usually take them on an extensive tour of the community, past the Intracoastal Waterway, through the immaculate Village with its staple businesses, by the waving grasses of the marshlands, past sparkling lagoons, stately stands of live oak, and neighborhoods full of elegant, well-kept homes. They tour this jewel of an island, and invariably have a similar response. They all say, "<u>now</u> I see why you moved to Savannah."

There was method to our madness after all.

COURSE DISCOURSE:
AN OVERVIEW OF THE LANDINGS

Plantation, number 9. Courtesy of The Landings

When my neighbors, golf companions and acquaintances find out I rate courses for a national publication, they invariably ask me, "which course at The Landings is the best?" A difficult question to be sure, as The Landings offers an embarrassment of riches. With six championship layouts even the most avid golfer can play to his heart's content, and not tread upon the same fairway twice in a given week.

I feel the best way to answer the question is not with an objective response of which courses are superior or inferior, but rather state which are the ones I personally prefer to play. I choose this tack for two reasons. The first is because one man's trash is another man's treasure. There's no need to impugn a design that might be a questioner's personal favorite, or speak glowingly about a course that he might deem better suited as a bird sanctuary or wildlife refuge. The second reason is that there are no objective answers anyway. The art of golf course rating is as subjective as restaurant, theater or movie reviews. Argue all you'd like, but there are no right or wrong answers in the end.

Palmetto, an Arthur Hills creation, is a course I never grow tired of. This 1985 design is a trophy wife of a golf course; beautiful, but exceedingly difficult. The shot-making requirements are practically unrelenting. The first four holes are demanding, the final three, particularly 16 and 17, are brutal. The par 4 penultimate hole is probably the single most difficult par to make of the 108 holes on the property. In between the tough start and furious finish are challenges of every sort. These include forced carries over marsh and water, well placed, cavernous fairway bunkers, multi-tiered greens, ball-swallowing grassy swales, and tilted fairways.

Why then, would anyone other than a masochist enjoy Palmetto? First, you need most every shot in the arsenal to play the course well. Secondly, the beauty of the layout, particularly on the eastern marsh holes on the inward nine, is almost unsurpassed on Skidaway Island. Palmetto is generally considered the most difficult of the six on the property. To my mind, it's also the best.

The Landings offers an embarrassment of riches. With six championship layouts even the most avid golfer can play to his heart's content, and not tread upon the same fairway twice in a given week.

Deer Creek is perhaps the most popular of the island's six courses, and for one reason in particular. This 1991 Tom Fazio design is laid upon a 120-acre plot, a larger piece of property than any of the others. Simply put, there's more room to drive the ball, and a sense of spaciousness on the golf course that's lacking on some of the older designs. The par 3s on Deer Creek, particularly on the outward nine, are fabulous tests, both challenging and lovely. The course also features a super little par 4, the drivable but problematic 5th. It's less than 300 yards from the championship markers, but the yawning lagoon flanking the fairway and fronting the green provide plenty of teeth. The closing holes on each nine are daunt-

ing, probably the two toughest finishers on the island. The course features extensive mounding framing the fairways on many holes. It's an artificial look, but it helps to keep offline shots in play, and offers a distinct visual characteristic. This feature, along with a number of elevated tee boxes, makes Deer Creek a uniquely enjoyable test among the offerings in the community.

A recent tweaking lengthened some of the par 3s and par 5s. The course isn't quite as score-friendly now, but it still captivates.

Oakridge was the encore delivered by Arthur Hills in 1988. While there's plenty of inherent difficulty, this is a journey with considerably less angst then his initial island creation of Palmetto some three years earlier. Fairways are generously cut, but so are bunkers, particularly the greenside devils situated on the first few holes. Many a medal score have been ruined less than 20 minutes from the clubhouse as frustrated players fail to extricate themselves from these diabolical sand pits with imposing sod walls. Hills chose to install a series of insidious low berms that serve to compromise select fairways. On holes 2, 4 and 17 in particular, a reasonable drive or approach can nestle against this low wall of thickly grown rough. It makes a pitchout the only viable alternative, when a ball that came to rest just a foot away could be advanced towards or onto the green. It seems unfair, but then again so is golf itself.

This course has the most interesting variety of greens on the island, both in shape and contour. All Landings veterans have been caught on the wrong side of the "dead elephant" buried on #6, while #14 is so narrow side-to-side it must be traversed in single file.

Oakridge also has the single most memorable hole on the island, the infamous cemetery hole, a gorgeous and historic risk/reward par 5. Bold players can make three, but run the risk of seven. Conservatives can plod their way, with no muffed shots, to a simple par. It's a great hole on a fine golf course.

Magnolia is an Arnold Palmer design, the first nine debuting in 1977, and the second nine opening some two years later. Before the extensive renovations performed in 2001, Magnolia was the most daunting shot-making challenge on the island. Many tee shots and certain approaches had to bend in the proper direction to avoid trees or wetlands, and find the best angle to the green. The new Magnolia has been lengthened, broadened and beautified, but some of the unique qualities of the initial design have been removed.

The new-look sandy waste areas filled with scrub plants afford an unusual perspective from the tee box. But only those players who can't reasonably expect to break a hundred will be intimidated by the minimal forced carries these new plantings require. It's a bit of sizzle, but not much steak. Much like Palmetto, Magnolia begins with a series of demanding holes before loosening its grip a bit midway through the outward journey. The back nine maintains much of the original character though, especially on holes 13 through 16. Placement becomes more important than power, bringing the shot-making values that once permeated much of the golf course back to the forefront.

Marshwood was the first course built on Skidaway Island. This Arnold Palmer creation debuted in 1974. One of Savannah's long-held nicknames is "The Beautiful Lady with a Dirty Face," which refers to the impressive architecture found in the historic district of town which was for many years in serious disrepair. The same nickname might be applied to Marshwood as well. The design is impressive and the vistas magnificent, but course conditions have deteriorated quite drastically in recent times. The total renovation project, scheduled for 2003, will hopefully return this initial design to its former glory.

Marshwood features dense stands of pine and oak which crowd landing areas, putting a premium on driving accuracy. Nowhere on the island will a player encounter as many par 4 doglegs. Only the tiny but treacherous 17th could be described as relatively straight-away, but even this hole requires a precise approach between bunkers and an encroaching tree. Several of the best holes have contoured putting surfaces which dramatically abut the eastern marsh, most notably the delicate 3rd, daunting 11th and dazzling 12th, perhaps the most scenic of the island's two dozen par 3 holes. Let's hope the new, improved Marshwood debuts with its unique character intact.

Plantation is a Willard Byrd design, and debuted in 1982. The opening two holes at the development's southernmost course are unquestionably the most difficult start on Skidaway. This is because the routing was reconfigured some three years after it opened, and this taxing twosome, originally numbers five and six, became the openers. There's plenty of difficulty on Plantation aside from these fierce par 4s, but much of the exertion necessary is mitigated by the inherent beauty of what is the single most scenic course on the island. The eighth through tenth in particular -- an uphill par 3, a tricky and reachable par 5 and a short but dangerous par 4, respectively -- are as striking a trio of golf holes as you'll see in this area, outside of the closing stretch at Harbour Town. It's not 7, 8 and 9 at Pebble Beach, but it's impressive nonetheless.

Every other course at The Landings, for that matter, virtually every other course in the Lowcountry, returns towards the clubhouse after nine before heading out again. Not so Plantation, which skirts the southeastern marshes at the furthest point on the property, and doesn't return to the clubhouse until after the figure eight routing reaches completion on the final hole. My only real quibble with the layout is that it features a few awkward par 4s. The fourth, seventh and sixteenth, the latter two in particular, require driving accuracy that is rarely seen on a consistent basis. You can call them unfair, or urge me to spend more time on the practice tee. Either argument is valid in my book.

I've been fortunate to have spent plenty of time at practically all of the Lowcountry's finest golf developments. The lush life at places like Belfair, Long Cove, Berkeley Hall, Haig Point and Colleton River is nothing short of extraordinary. The courses are all stunners, with impressive practice facilities and immaculate conditioning. Course pressure is rarely an issue, tee times a cinch to obtain and service beyond reproach. But in all honesty, I would never trade my Landings membership for any of them, one to one.

Basketball coaches like to say "you can't teach height," and in the same vein, there's just no substitute for the amazing diversity of golf experiences found on Skidaway Island. The closest I've seen is Desert Mountain in Scottsdale, which features five Jack Nicklaus designs with a sixth course in the offing. But spectacular as they are, there's a certain sameness to these Arizona layouts. The Landings, with six fine courses envisioned by four different architects, offers a unique, some might go so far as to say unrivaled golf experience. For those who seek the ultimate in both quantity and variety in a private club setting, The Landings Club remains the bellwether.

PERSONALITY:
DOUG HANZEL – DR. DYNAMITE

Do you attribute Dr. Doug Hanzel's golf excellence to the fact he's been playing for more than 40 years? Perhaps, but the more pertinent question is how does a guy in his mid 40s already have such a long history in the game? While we're asking perplexing questions, try this one: If you're as slim as a 2-iron and lighter than a graphite shaft, how do you regularly drive the ball about 270 yards, seemingly with little effort?

The answer to both questions comes down to a single word. Timing. Hanzel, following in the footsteps of his dad and older brothers, took his first swings at the age of three. Now more than four decades later, he maintains a "plus" handicap though he's almost entirely self-taught. He's a 5'7" welterweight, but fundamentals and club control allow him to swing at 80% of capacity, yet routinely out-drive his playing partners, most of whom outweigh him by 50 pounds or more. Clearly, there's more to this pulmonary critical care physician than meets the eye.

Tad Sanders is the Director of Golf at the Landings Club on Skidaway Island, and has borne witness to Hanzel's ten club champi-

onships in the last eleven years, including the last eight in a row. "Dr. Hanzel is one of the most consistent ball strikers I've ever seen, and really manages his game beautifully around the golf course. He's also one of the greatest competitors I've seen because he's never satisfied, he's always looking to improve."

Hanzel has lived in Savannah since 1989, but was raised in a Cleveland, Ohio, suburb where there was plenty of room to spread out and swing a golf club. "We had three acres of grass around our house, and there was a country club nearby," begins the doctor. "My dad loved the game, and my brothers and I were caddies and all learned to play. That's how it all started." The Hanzel boys were all single digit-players, but Doug was clearly the most talented. He played on the golf team throughout all four years in high school, and won the Ohio schoolboy championship as a senior in 1974. He matriculated at Kent State in 1975, some five years after the infamous shootings at the beginning of the decade. "Even though the incident had been several years earlier, there was still lots of sentiment and memorials on campus. But the protesting was long gone by the time I arrived."

The success continued at the collegiate level, including all-conference designation three times. He tangled with future pros like John Cook, Gary Hallberg and Mark O'Meara during these years. He claims he held his own in tournaments, but could see they were a cut above the average college golfer. "You could tell these guys had an edge over most of the players out there," states Hanzel.

There were two major factors that kept him from pursuing a pro career. "I guess any good player thinks about what he could do as a pro, but I was influenced by my eldest brother David. He's ten years older than me and was a physician. I wanted to be one as well. I also didn't like to travel that much, and wasn't interested in a life on the road."

Hanzel never so much as took a step in that direction. He applied to medical school instead of Qualifying School, and chose Wright State in Dayton, Ohio, for his medical training. It's been more than 20 years since he made that decision. There are no regrets, but he admits he occasionally wonders what might have been down the other fork in the road. "In my heart I truly believe I could've made it out there, especially when I think of channeling all the energy I had put into my studies into my golf game instead. I feel I would've had a legitimate chance at making a living out there."

Instead he's played an impressive amateur schedule over the ensuing years. Hanzel debuted in the U.S. Amateur Championship in 1978, the same year his college buddy John Cook beat Scott Hoch in the finals. Hanzel beat the New York City Amateur Champ in the first round before losing in the

second round. He's qualified seven times since, including the 2001 tournament at East Lake in Atlanta. His best showing was in 1996 when he finished 15th in the preliminary rounds, but lost his first round match to Trip Kuehne, the man who took Tiger Woods to the wire before losing in the Amateur final in 1994.

If you're as slim as a 2-iron and lighter than a graphite shaft, how do you regularly drive the ball about 270 yards, seemingly with little effort?

He's made quite an impact in local events over the years as well. Hanzel has won the Savannah City Amateur Championship, is a former winner of the LaVida Championship and has won the Oglethorpe Invitational three times. Although there's plenty of competition to be found up on Hilton Head, the doctor is generally too busy with either family or his medical practice to compete in South Carolina. "I've got a son and daughter, and both are teenagers. They keep me busy, and really limit the amount of time I spend around the course."

Don't look for Hanzel to emulate another long-time Georgia amateur standout, LaGrange's Allen Doyle. You'll recall that the driving range owner turned pro in his late 40s, in the hopes of a successful Senior Tour career. His plan worked spectacularly, with seven wins and more than seven million dollars in career earnings. Of course one of the major differences is Doyle made about $30,000 a year at his range, while Hanzel is a successful doctor in private practice. Even though his kids Katie and Drew will be off at college by the time he turns 50, he's not planning an attempt at the pro ranks. "I'm going to stay an amateur," concludes Hanzel. "My hope is to start playing some high level senior amateur competitions when I turn 55. My medical practice is pretty intense. It's intensive care work with very sick people. I deal with respiratory failure, ventilators, and many patients just don't make it. I'll be ready to ease up in another ten years or so."

Hanzel's logical move at that stage will be into the Society of Seniors, the nation's most prestigious and selective senior golf association. Membership requirements include USGA competitive experience and a club championship in an applicant's background. Needless to say, the good doctor won't exactly be sweating out the screening process.

For now though, he's content to enjoy the kids, keep shooting rounds of par or better, and playing the occasional championship. His son Drew is starting to get the golf bug, and father and son are spending more time together on the course. If the offspring shows even a fraction of the desire and talent his dad has, the next generation of Landings golfers will be as frustrated as the current. Doug has his name on the trophy ten times, and figures to be good for at least a half dozen more. If Drew follows suit once he comes of age, there's no reason the tournament trophy shouldn't undergo a name change. Never mind the unwieldy handle Landings Club Championship. Let's just call it what it is. The Hanzel Cup.

PERSONALITY:
J.D. TURNER – PREMIERE TEACHING PRO

Imagine the prestige in being one of the top 100 practitioners in your chosen field. If you were a pro athlete, there'd be a fleet of luxury imports and a multi-year contract. An architect would produce soaring monuments that attest to his distinguished position; a writer might have a Pulitzer Prize, a multi-book publishing contract, or both.

Savannah's J.D. Turner knows what it feels like to be among the best. The Iowa native and current Landings resident has been one of GOLF Magazine's Top 100 Instructors since the list first debuted more than a decade ago. Initially the list was limited to the finest 50 instructors in the land, so clearly Turner has been at the top of his profession for a long while.

"I'm from a small railroad town called Perry, Iowa," begins Turner. "I spent my entire career in Iowa until moving to Savannah in 1997." There were two main reasons that Turner moved from the Corn Belt to the coastline. The first was a friendship with former Iowa State basketball coach Johnny Orr, at the time a Landings resident. They became friendly at the Des Moines Country Club,

where Turner served as Director of Golf for 14 years. J.D.'s wife Brenda is an accomplished golf artist, and was commissioned to produce paintings and prints of the Deer Creek golf course at The Landings prior to its opening. "We liked what we saw when we came up for Brenda's art project, and that's how we ended up relocating here."

After more than 25 years as a club pro, the PGA Master Professional started his own company in 1994. The J.D. Turner Golf Group's principle business is running corporate golf schools for blue chip clients like Monsanto, Martin-Marietta, Gatorade and AC-Delco at various locations throughout the country. He employs six or eight instructors at a time from a rotating group of 20-30 handpicked PGA pros. They run the specific clinics based on their schedule availability, the geographical location of the school and other factors. "My staff is comprised of colleagues I've met over time, terrific teachers, wonderful players and great communicators. I know guys from the north whose clubs shut down in the fall and are available for winter schools. I have colleagues in Florida who want to teach in summer during their slow period, so it all works out nicely."

Turner, a man in his early sixties, came to his profession quite naturally. His father was a teacher and athletic coach, his mother a fine golfer. Growing up in rural Iowa during the polio years of the '40s and '50s, there was no public swimming allowed. There was no little league or tennis, but there was a nine-hole town golf course. "It was nothing for my friends and me to go round and round, 36, even 45 holes a day in summer," he recalls.

Turner captained the University of Iowa golf team, and while he didn't quite measure up to his Ohio State contemporaries Jack Nicklaus and Tom Weiskopf, he did have enough game to become a member of the All Big 10 Golf Team. Later in life after turning pro he captured five Iowa Open titles. Still later he competed in two U.S. Senior Opens and a Senior PGA Championship. But it was his teaching, particularly his teaching innovations, which paved his way to prominence.

Decades before the advent of The Golf Channel and their ubiquitous instructional programming, Turner pioneered a statewide cable program called the *Iowa Golf Show*, later renamed *J.D. Turner Golf University*. "I was a high school teacher for several years before I became a golf professional," recalls Turner. "I recognized that many of the high school golf coaches in the state were biology teachers, math teachers and the like, and probably didn't know all that much about coaching or teaching golf. I distilled the TV show archives down to the essentials, and created a pair of 3-hour videotapes." Turner's total package was geared to Phys. Ed. departments, as well as high school and college golf

teams, and eventually reached almost a million coaches, teachers and golf students in the nation. It included 30 lesson plans for effective practice, a Q&A directed towards golf as taught in a classroom environment and specific fixes for faults in the swing. It was this unique approach to instruction which brought Turner to the attention of the GOLF Magazine selection committee in the first place.

Decades before the advent of The Golf Channel and their ubiquitous instructional programming, Turner pioneered a statewide cable program called the Iowa Golf Show, *later renamed* J.D. Turner Golf University.

"It's opened many doors," concludes Turner, speaking of his official designation by one of the game's premiere publications. "At the time, it even got me a small raise back at my country club!"

Turner's teaching philosophy is quite simple. "80% of the battle comes before you take the club back," he states earnestly. "If a player has solid pre-swing fundamentals, if their grip, stance, arms and golf ball are all in the proper position, then most folks will be able to learn to hit it pretty well. You have to be a superb athlete with great hand/eye coordination to constantly make the necessary compensations in your swing to make up for poor fundamentals."

In addition to his *GOLF Magazine* designation, Turner has been honored as the youngest member inducted into the Iowa Golf Hall of Fame, and has been a finalist on numerous occasions for the National PGA Teacher of the Year, although he has yet to bag the top honor. "I'm starting to feel like Susan Lucci," he says with a laugh.

Although their roots run deep back in Iowa, the Turners are extremely satisfied with their move to Savannah. "The weather is great, the airport makes it easy for me to get to my golf schools, Savannah College of Art and Design appeals to Brenda, and as a former history major, I love that aspect of the city. It's really a good fit for us."

COURSE DISCOURSE: THE WOODYARD

Number 13. Courtesy of The Woodyard

Savannah will never offer the quantity or variety of golf experiences that are found on Hilton Head. But the golf scene has changed dramatically across the Georgia state line in recent years, and several worthwhile venues have debuted that rival some of the upscale resort facilities around the island. Not the least of these upstarts is The Woodyard, an impressive Bob Cupp design less than five minutes from the Savannah Airport off of Interstate 16, perhaps a 40-minute drive from the Hilton Head bridge.

As the name implies, The Woodyard is cut from the thick pine forests west of the city. In the years to come it will be a centerpiece of a

nascent housing development known as Savannah Quarters, but for the foreseeable future it should remain a secluded and rustic walk through the woods. This is a shot-maker's golf course, giving the appearance of wide open fairways that are occasionally bordered or bisected by near-hidden hazards. It's a stern test of the game from the championship markers, measuring a shade less than 7,200 yards. The penultimate tees are no pushover either at 6,700 yards, with a course rating of 73.5 and a slope rating of 132.

The opening nine is distinctive because the counter-clockwise routing features three par 3s, 4s and 5s. The back nine is laid out in a clockwise direction, with a pair of one-shot and three-shot holes offsetting five par 4s.

Average golfers tend to favor par 3s and 5s, as they often represent the most legitimate opportunities to make par. This is not necessarily so at The Woodyard though, particularly on the outward stretch. The 4th and 6th are one-shot holes that average more than 200 yards in length. The eighth is only a mid - to short iron at about 150 yards, but bordered tightly by water right and an insidious and almost invisible lateral hazard to the left.

Average golfers tend to favor par 3s and 5s, as they often represent the most legitimate opportunities to make par.

The par 5s are sturdy in their own right. They average almost 540 yards in length, meaning the duffed shot one can occasionally recover from on many three-shot holes will make par an unlikely result here. All three are fairly severe doglegs with plenty of water, wetlands and waste bunkers flanking the lengths of the fairways. Landing areas are mostly generous from the tee box, and solid shots will stay safely in play. But errant approaches that drift offline or fail to carry the variety of hazards will quickly add penalty shots.

There are some par 4s with real teeth on the inward journey. The eleventh is a bit gimmicky; an awkward 270-yard hole where most folks can't hope to drive the green, but hit it plenty far enough to find trouble elsewhere. Other than that aberration, the remaining quartet are long and strong, aver-

aging almost 430 yards each with no shortage of sand and water influencing the line of play.

The Woodyard has some of the most unusually shaped and difficult greens in the region. Some are extremely narrow, and others have offshoots and protuberances where an evil superintendent can stick a flagstick that would necessitate a laser guided approach to come close. Still others have curious swales and runoffs, which will cause poorly struck approach putts or chips to meander well away from the intended line. In summation, it's a golf course that's demanding on a variety of levels.

The Woodyard is managed by Troon Golf, which is one of the premiere names in upscale golf management. Course conditions are generally quite good, particularly the putting surfaces. This is a course that will challenge but not overwhelm the handicap player, provided he chooses the appropriate tee box from the five choices available. It won't be too long before the course hosts a major competition of some kind though, as the length, omnipresent hazards and variety of diabolical pin placements available on the greens make it a suitable test for the highly skilled player as well.

ESSAY: GOLF BOOKS WE HOPE TO NEVER SEE

Father's Day is fast approaching, marking the apex of the golf book-buying season. Every year hundreds of new titles are introduced; instructional, mystical, melodramatic and humorous, many penned by celebrated players. This year's crop of "name" authors is a bit thin, but here is an advanced list of what you'll be seeing on bookstore shelves next season. It makes us wonder; do we really have to clear-cut thousands of forested acres for this crap?

"The Corrections" – 25 Ways to Grip and Re-grip the Golf Club
by Sergio Garcia

"Long Day's Journey into Night" – How to Stay Focused During a 6-Hour Round
by PGA Tour winner Glen "All" Day, foreword by Bernhard "Linger" Langer

"To Kill A Mockingbird" – The Strangest Golf Shots I've Ever Hit
by Tom Kite

"The Daly Double" – How Two Fluke Major Championships Changed My Life
by John Daly, foreword by Andy North

"Hillbilly Name, Plenty of Game!" – How We Play Golf
by Billy Ray Brown and Jay Don Blake, preface by Billy Bob Thornton

"I, Stank" – the autobiography of Paul Stankowski

"For Whom the Bell Tolles" – The Rise and Fall of a Tour Player
by Tommy Tolles

"The Kitchen Cink" – Favorite Recipes for Golfers
by Stewart Cink, introduction by Edward Fryatt

Couples and Love on *"Couples in Love – How Golf Can Make Your Marriage Stronger"*
afterword by Hal Sutton

"Watch out for the Duck! (hook)" – Golf and Bird Watching
by Ian Baker-Finch, forward by Tim Herron, afterward by Jonathan Byrd

"The Glasson Menagerie" – Bill Glasson's Guide to Pet Care

"Holy Toledo!" – My Spiritual Awakening Through Golf
by Esteban Toledo

"The Wizardry of Westwood" – Lee's Guide to Great Golf
preface by John Wooden

"Don't Call me Thurston!" – the autobiography of Charles Howell III

"Alternative Lifestyles in Golf"
by Brian Gay

"The Harder your Name, The Stronger your Game" – Life as a Top Female Amateur
by Jenny Chuasiriporn, Vida Nirapathpongporn and the Wongluekiet twins

"Golf is not a Game of Giants"
by Bob Costas and Peter Kostis, introduction by Ian Woosnam, afterword by Rory Sabbatini

"Pressing the Flesch" – How Golf can Help you in the Business World
by Steve Flesch

"Hogan's Five Fundamentals" – A Guide to Making Vapid Hollywood Sequels
by Paul Hogan, preface by Sylvester Stallone

"Golf Course Rap"
by Vijay Singh with Charlie Rymer

"High School Confidential – Proms, Pimples and Peachfuzz"
by Ty Tryon

"Goydos!"
by Paul Goydos

"Grateful Fairways"
by Bob Weir with Mike Weir

"A Man of the Peoples"
by David Peoples

"Lady's Man: The Frank Lickliter Story"
by Frank Lickliter

"Bjorn to be Wild"
by Thomas Bjorn

"I am the Walrus"
by Craig Stadler, preface by Ringo Starr

"Oh, Kanada! – The Memoir of Craig Kanada"

"Tunics on the Tee Box" – How to Dress for Success and Win Every Major Championship
(except the U.S. Open)
by Nick Price and Nick Faldo

PERSONALITY:
BILL DEGENHART – CAPTAIN CONNECTION

There's no way of gauging exactly how many avid golfers there are in Savannah and the Carolina Lowcountry, but it's safe to assume that the number is well into the tens of thousands. Dr. Bill Degenhart is one of these golfers. The Savannah ophthalmologist is a fine player, albeit not of championship caliber. There are many area residents who've been playing longer than Degenhart, or play more often, or with more skill. But it would be hard to find an example of a more enthusiastic player, someone who loves golf more, or any non-professional who's experienced the sheer variety and quality of courses that Degenhart has in his many golf adventures. The man is in his early 50s, but has already lived out two lifetimes of golf fantasies.

Born in Buffalo, Degenhart had the foresight to escape the cruel New York winters at the tender age of six weeks, and migrated with his parents to the more inviting climate of Florence, South Carolina. His golf and schooling commenced simultaneously at age five, and he excelled at both. He was a product of the South Carolina educational system

all the way through medical school, and at times carried a handicap as low as one. For the first 20 years of his golf life, Bill Degenhart was no more connected than anyone else. All that changed when he moved to Philadelphia in the late 1970s for his medical residency. He chose the Wills Eye Hospital because of its reputation as one of the finest facilities in the world, little realizing he would soon lay the groundwork for a similar affiliation with a golf club of equal renown.

"My first year at Wills, I was invited to play at Merion (#7 on the Golfweek Magazine list of America's Best Classical courses) by a fellow named Mark Porter, who's mom Dorothy was the reigning U.S. Senior Amateur Champion. Mark eventually put me up for membership at a club in the city called Riverton. I had grown up playing at the Florence Country Club, so it seemed like a natural thing to do." Riverton was a pleasant club, but there were much bigger things in the works for Degenhart. "In 1980, I was asked to play in a doctors' outing at Pine Valley (#1 on the Golfweek Magazine list). I played pretty well, shooting in the 70s, and I guess it impressed my host. He said, 'You ought to be a member here.' He introduced me to some other members, and I was put up for membership in 1984. There was a waiting list at the time, and I was eventually accepted into the club in 1991." Former Savannah resident Bill Goldthorpe was instrumental in helping Degenhart gain admittance into this most coveted of clubs. Goldthorpe, who left the Lowcountry some years ago for the Raleigh-Durham area, was a longstanding friend of Pine Valley Chairman of the Board, Ernie Ransome. "If Bill hadn't vouched for me, I don't know if I would have been invited to become a member," admits Degenhart.

"When I'm around a table having drinks, and I'm the only guy in the group without a private jet, then I feel like I'm out of my league."

In the eleven-year period between his initial exposure and acceptance into Pine Valley, there were major changes occurring in the lives of Bill and Jan Degenhart. They decided to settle in Savannah in 1981, and initially chose to live on Skidaway Island. "The truth of the matter is that I drove into The Landings, which at the time only had two golf courses, and decided that it would be a fine place to live," recalls the eye doctor. He built a multi-specialty practice, now known as the Georgia Eye Institute, and eventually added 13 partners. The Degen-

harts also built a family. Their oldest son Jay is an accomplished golfer, and attends the University of North Carolina, as does his younger sister Leanna. Their youngest son Blake still lives at home.

In the years prior to joining Pine Valley, Degenhart continued to develop golf contacts throughout the country. A weeklong excursion more than fifteen years ago with seven of his friends stands out as particularly memorable. Normally, one is given the choice between quality or quantity, but with the good doctor acting as travel agent, both were available in abundance. "We began by playing 36 holes at Saucon Valley in Pennsylvania the first day (#82 on the Golfweek list), then played 18 at Merion the day after, and then 36 at Pine Valley the day after that. Then we moved on to Long Island and played 18 at National Golf Links of America the next morning, (#11), followed by 18 at Maidstone (#33) in the afternoon. The next day we stayed on Long Island, playing 18 at Shinnecock (#5) and then 18 more in the afternoon at Garden City (#18). We moved on to Sleepy Hollow in Westchester County, followed by 18 at Mountain Ridge in New Jersey. The last day, we played 36 at Baltusrol (#39)." It was a trip he'll never forget, but one he won't repeat. "It was a great time, but a logistical nightmare," recalls Degenhart. "In most cases, we had to play with at least one member at each club, and it was tough to coordinate with everyone. I would have to tell our prospective host, 'We're looking forward to playing, but we have to be finished by 1:00 so we can make our next tee time.' The trip was super, but I wouldn't attempt it again."

In addition to his membership at Pine Valley, Degenhart is also a founding member of Secession Club; the caddies-only gem in Beaufort, and the Ogeechee Golf Club at the Ford Plantation in Richmond Hill. "Of all the places I play, Secession is probably the most enjoyable. There's less pretense involved there than anywhere else. I joined the Ford Plantation because I used to play down there in the early '80s with John Gustin, the old head pro, so I knew what a great course it was. When I had the opportunity to become a member, I couldn't resist."

Through circumstances and serendipity, Bill Degenhart moves easily through the privileged and rarified air of the nation's premiere golf clubs, but it doesn't really faze him. "When I'm around a table having drinks, and I'm the only guy in the group without a private jet, then I feel like I'm out of my league," he offers with a grin. "On the other hand, if I had the low score on the course that day, which is sometimes the case, then they might feel out of their league. The best part of all of my trips and golf experiences is really the camaraderie," concludes Degenhart. "The courses are wonderful, obviously, and the competition is great also. But it's really the people, the stories and the memories of good times that I cherish the most."

Talking golf with this unassuming eye doctor leads to a single and simple conclusion. There are only two types of golfers in greater Savannah. There are those with the contacts and connections of a man like Bill Degenhart, and those that wish they had the same.

Dr. Degenhart's Diamond Dozen:
Pine Valley-NJ
Cypress Point-CA
Augusta National-GA
Shinnecock Hills-NY
Merion-PA
San Francisco Golf Club-CA
Saucon Valley-PA
Winged Foot-West-NY
Crystal Downs-MI
Ballybunion-Ireland
Garden City-NY
Seminole-FL

ESSAY: GOLF TIP – KEEP YOUR HEAD DOWN, AND YOUR MOUTH SHUT!

There are lots of annoying things about golf; five hour rounds, plaid pants and three putt greens chief among them. To this list I add another vexation, one that is seldom mentioned but an irritant nonetheless. I can't stand it when other players comment prematurely on your shots. Let me cite three hypothetical examples; one each from the tee box, fairway, and green.

You miss-hit your drive. It leaves the clubface O.K., but then starts fading faster than a Kevorkian patient. You're watching it career wildly to the right, and this is what you hear: "Nice drive! Oh oh, hang on...hang on...is that OB over there?"

You don't quite get the entire club on an approach shot. You hit it kind of chunky, and though you harbor hope it might miraculously reach the green, deep down you know it's headed for disaster. This is what you hear: "It's right at the stick! Great shot! Oh man, I can't believe that didn't carry the pond. Maybe there's wind up there."

You hit your putt with the touch of a Bulgarian stonemason; you know its way too strong. As the ball steams too quickly towards the hole, you hear either "It's in!" (it's not) or "Nice putt." The comment comes as the ball just misses the hole, but before it comes to rest eight or ten feet by.

Call it trivial, label me as sensitive, I don't care. My feeling is that you are bothered enough on the course when you make stupid mistakes that cost you strokes. Having some dopey optimist in the group dispensing ill-timed praise only compounds the frustration. At least it does to me. I am not advocating silence on the golf course, far from it. I am gregarious by nature, and am rarely qui-et for more than a minute at a time. Compliments, commiseration and needling are all integral parts of the whole golf experience. My point of contention is solely based on the timing of these comments. If you are going to talk about someone's good shot, make sure it is actually good. Otherwise you look kind of silly.

There is another side of this coin as well. Though I am rarely guilty of cockeyed optimism regarding my companions' shots, I will often comment negatively about my own shot before it comes to rest. We have all done it. Maybe you scull a shot that ends up next to the flag. Perhaps you admonish yourself by saying "hit it" when you perceive a putt has been babied, only to watch it make its way to the hole and drop in.

You miss-hit your drive. It leaves the clubface O.K., but then starts fading faster than a Kevorkian patient. You're watching it career wildly to the right, and this is what you hear: "Nice drive! Oh oh, hang on...hang on...is that OB over there?"

A wise golfer I once met summed it up nicely. Don't let your mouth ruin a good shot.

I know a man up north named Sirkin. He's a nice enough guy, but one with a habit on the golf course so annoying it is actually laughable. Ten or twelve times a round he will exclaim, "It's in the hole!" while watching your putt. It may be a forty-footer, and the ball is only halfway there, but he is sure it's going to fall in. My friends and I call it "Sirkin-cising," as in, "Don't Sirkincise me, that shot was terrible!"

Let me suggest a more generic description. A one sentence maxim that should be as integral a part of golf etiquette as raking a bunker and keeping clear of someone's putting line. Do not comment on a golf ball in motion. Simple, fail-safe, and all-encompassing.

There is all the time in the world to praise, laugh at or sympathize with anyone in your group. Just give it that extra few seconds so you know exactly what it is you're talking about.

COURSE DISCOURSE: THE FORD PLANTATION

Although its history as a plantation can be traced to the 1700s, the modern era of The Ford Plantation can be described in three segments. In 1935 automobile magnate Henry Ford purchased 70,000 acres on the Ogeechee River, and constructed an impressive manor home. In 1982 wealthy Saudi Arabian entrepreneur Ghaith Pharon purchased the manor home and 1,900 acres. He set about customizing the grounds into a private playground, and hired Pete Dye to design a golf course, completed in 1985. Pharon left the country several years later and the course and grounds fell into disrepair. Developers Chip Dolan, Peter Pollak

and Steve Schram purchased the property in 1998 and began refurbishing it. Their hope was to turn the Ford Plantation into one of the nation's most desirable second home retreats, and they've done exactly that.

The developers don't describe The Ford Plantation as a golf community, but few would argue that the Dye golf course isn't a centerpiece attraction. Horseback riding, shooting, boating and fishing are part of the amenities package, but the Ogeechee Golf Club is arguably the single biggest draw.

Of the three major elements that combine to make the course exceptional, the most impressive is the absolute lack of play, which is the rule and not the exception. Assistant pro Andy England provides the details of a facility that sees less play on any given day than most courses see in an hour. "In our busy season, October through May, we average about 25 players a day. In the summer we might only see one or two foursomes a day." The 40ish England has a facetious explanation for his prematurely gray hair. "I really like this job, but with all the traffic through the pro shop it can get a little stressful."

The second notable factor is the dramatic difference between the two nines. Taken as a whole the course is extremely spacious. It was constructed on more than 260 acres, 120 of which are sodded, with almost 250 additional acres of freshwater lakes. The outward nine holes are a straightforward journey through forests of pine, oak and palmettos. With expansive waste areas festooned with ball-eating love grass, and smallish greens averaging 4,000 square feet, shot making is at a premium. There are an abundance of bunkers in all shapes and sizes, and water in play on six holes. This part of the course is compelling but conventional, while the back nine offers stark contrast.

Dye constructed the inward nine holes in the midst of the plantation's antebellum rice fields, which Henry Ford had turned into lettuce fields in the 1930s. This nine is reminiscent of a Scottish links, bereft of trees, with incidental water intruding on just a few occasions. The designer moved more than two million cubic yards of earth during construction, with striking results. The middle portion of the final nine holes is of particular interest. Players strike their drives next to Lake Clara, named for Ford's wife. The tee boxes are just steps from the lakeside, with all sorts of fish and fowl close at hand. The wildlife's proximity is marvelous, the setting both pristine and slightly primeval.

The third element of note is the demanding nature of the golf course, which stretches to almost 7,100 yards in length, and carries a slope rating near 140. The only folks who belong on the championship tee box are either long hitters with low single-digit handicaps, or the grounds crew. Everyone else should take a few giant steps forward.

Interestingly, Dye deigned to keep all of the par 3s under 200 yards in length, and the par 5s well under 600 yards. He saved his venom for the par 4s, which average almost 415 yards each. Discounting the two shortest, including a drive and pitch hole, and the remaining par 4s average more than 430 yards each. When the wind blows, as it often does on this virtually treeless back nine, then players are confronted with a golf challenge as tough as anything on the southeast coast.

The architect himself has been quoted as referring to the Ogeechee Golf Club as "arguably my finest southern design." A strong sentiment, considering classics like Harbour Town, Long Cove and the Stadium Course at TPC Sawgrass are within close proximity. Dye might be overstating a bit, but it's hard to argue that his work at The Ford Plantation, little played and to this point little known, isn't deserving of some serious attention.

ESSAY: A PAIR OF AWARD-WINNING GOLF POEMS

With Valentine's Day fast approaching, I was thumbing through one of my favorite poetry books recently, a marvelous compilation called "Golf, Love, and other Great four-letter Words." I found a pair of hidden treasures within the pages, two poems whose poignancy of subject, elegant simplicity of iambic pentameter, vivid use of metaphor and inspired rhyming pattern struck me as worth sharing. Mostly though, I was impressed with the author's lineage.

Elmer Kilmer was named Poet Laureate of Sandusky, Ohio, in 1996. A welder and part-time butcher by profession, Kilmer is the illegitimate son of the ex-wife of former Washington Redskins quarterback Billy Kilmer. He is a step-cousin three times removed to Hollywood heavyweight Val Kilmer, and most importantly, the great-grand nephew of famed poet Joyce Kilmer, who died in 1918. The first poem that follows won the silver medal in the welder-butcher division of the annual Northern Ohio Poetry Competition, Lake Erie Division back in 1999. The second effort, obviously written by a more introspective, some might say morose Kilmer, took honorable mention the following year in the same category. I'm sure you'll enjoy them both as much as I did.

"TEES"
(with Apologies to my Great-Great Uncle Joyce Kilmer)

I think that I shall never see a poem as lovely as a tee.
This simple peg, this piece of wood, precedes a pastime understood
by countless millions, lucky souls, who know the joy of eighteen holes.

You wager, pulling ball from pocket. Take your stance and launch your rocket.
Chase it, launch it, chase it down, until that ball rests underground.
Then take your tee and start again. This sport for gods, for kings, for men.

You may prefer to surf or garden, thinking golf makes arteries harden.
Shoot pool, lift weights, play bridge instead. Then you're among the living dead.

The living dead don't understand the thrill of blasting from the sand;
the ball descending like a feather, landing stiff, within the leather.
A dead-on chip that trickles in; for just one beat: Phil Mickelson!

Under pressure, a perfect drive! God it's great to be alive!
A fresh mown fairway, brand new glove, some would golf and not make love.
Am I off-base, think I've drank a few? Golf lasts half the day—can you?

So you're a watcher? Not a doer? Perfect! You can be a viewer!
Grab a snack and find the clicker: Daly, Duval, Lehman, Stricker, Tiger, Leonard, Els and Price;
Johnny Miller gives advice. The thrill of watching golf's elite—convenience; never leave your seat!

So stalk the fairways, grab a buggy, weather crisp or hot and muggy.
Switch to soft spikes, change your grips; pay your green fees, take your rips. Drive with power,
putt with grace; practice, pray, then make an ace!
My friend you'll know that life's begun, when they toast your hole-in one!

Count all your strokes and don't play slow. You'll break 90 before you know. Take some lessons, learn the swing. You'll quickly know if golf's your thing. And if it's not, then that's O.K. The course is too crowded anyway....

"PENCIL PUSHERS"

Of all golf's high-tech gadgetry
The one thing people rarely see
That has the power to change the game
Is the tool that's known by a household name.
For lower scores, master one utensil,
The unpretentious golf pencil.
On the cart, or behind the ear
A well-used lead can engineer
A scorecard one displays with pride
Instead of one you rush to hide.
This wooden marvel, short and stubby,
Used with care by wife or hubby,
Effortlessly will transform
A score much higher than the norm.
Sometimes a player must connive
Reducing seven down to five.
It's not a virtue to extol,
But it might just help you win the hole.
Please be judicious, sons and daughters
Pigs get fat, but hogs get slaughtered.
A four from five, or a three from four,
Just makes you a scorecard whore.
Don't be caught in gross misuse,
Never "shave" a stroke from deuce.
A word of warning to end this ditty,
I'll describe a scene that wasn't pretty.
A psychopathic simpleton
Once claimed he made a hole-in-one.
This bald-faced act was his downfall,
He thought his pals would not recall
He was banished, in disgrace…
He made a two, and claimed an ace!

PERSONALITY:
JACK CASON – AWESOME OCTOGENARIAN

I would like to extend belated birthday greetings to an occasional golf partner of mine. Jack Cason turned 80 not long ago. He celebrated in style, a smooth 81 with three birdies, and then a couple of beers. I don't know what impresses me about Jack more, his perpetually pleasant disposition, or his solid all-around golf game. He has not only turned four score, but he's still capable of turning in a good score.

Mr. Cason is a fairly new resident on Skidaway Island; he and his wife Jackie moved out to The Landings just a few years ago. This is not to imply that he's just getting to know the courses, though. Jack has been in Savannah since 1956 and a member of the Landings Club for almost thirty years.

Jack and I both play golf like men thirty years younger. He's got the strength of a fifty year old, and I often play like I'm in grade school. Many older golfers who are still capable of shooting respectable scores play a finesse game; bunting the ball around the course, staying out of trouble, and hoping to get up and down to save par. Jack's not like that. He still has the sem-

blance of a power game.

Nobody has ever confused me with John Daly, but I'm not particularly short off of the tee. I'm relatively young, and have reasonable coordination, so there's no reason I should be. When I catch a drive, and when Jack catches a drive, he's usually not more than a dozen paces behind me. At first this was frustrating, until I came to the realization that this man is simply exceptional. Analyzing somebody's golf swing is the last thing I'm capable of. But others have remarked that the reason Jack has maintained so much of his distance when other men his age have either given up the game or relegated themselves to the ladies tees, is that he fires his weight to the left side of his body on every swing. Driver through wedge, he has a consistent weight shift that gets most of his power into each shot.

In certain ways the man is definitely a throwback. He brandishes a hickory shafted putter that looks like it has the loft of a three iron. No matter. He misses those testy 3 and 4 footers about as often as he fails to address everyone in the vicinity as "partner," which is to say almost never.

Jack Cason's golf game of choice is contested from the blue tees. He doesn't play the championship tees anymore, but instead chooses a solid, mid-length course in the neighborhood of 6,300-6,400 yards. That's a commendable decision for any newly-minted octogenarian, but doubly so considering I'm acquainted with a number of men young enough to be Jack's sons, who wouldn't follow suit. They can't scurry to the forward tees quickly enough, as if playing the shortest and least challenging golf course available is some sort of birthright.

Jack and I both play golf like men thirty years younger. He's got the strength of a fifty year old, and I often play like I'm in grade school.

As a sports-minded youth raised in New England, I wanted to grow up to be like Yaz, or Bobby Orr, or John Havlicek. Time has passed, and I'm grown up now. Well at least I'm older. It's too late for me to grow up to be like someone else, but there is one thing I've become pretty sure of. When I grow older, I want to be like Jack Cason.

COURSE DISCOURSE: SOUTHBRIDGE

Number 8. Photo by Michael D. Ayres

It's been quite a while since an architect held the spotlight like Rees Jones did throughout Georgia in August of 2001. His renovation jobs at Atlanta Athletic Club and East Lake were the sites of both the PGA Championship and U.S. Amateur respectively, and his original design Ocean Forest hosted the Walker Cup down on Sea Island. Not a bad hat trick, especially with all three events playing out over a single 15-day period.

Jones has left his mark in the Atlanta area to be sure; not just his renovation work, but also the highly regarded Piedmont Driving Club, and his fine Oconee Course at Reynolds Plantation. It's in the eastern part of the state though, near the coastline and through the Georgia and South Carolina Lowcountry where Jones has been most prolific. Haig Point on South Carolina's Dau-

fuskie Island is another stalwart, practically in the same league as Ocean Forest. He's created several other courses around Hilton Head, and one little-known treat in Savannah, called Southbridge Golf Club.

The course at Southbridge is the centerpiece of a thousand-acre residential community just moments off of Interstate 16, less than 15 minutes from downtown Savannah. Carved out of wetlands and heavy forest, Southbridge is a course that demands accuracy on both tee shots and approaches. Water comes into play at least incidentally on all 18 holes, but in some cases it would take a mistake of epic proportion to actually find liquid.

Because the property has a tendency to hold groundwater, this is often a very soft and saturated golf course. Only strong players, not necessarily low handicaps but physically strong, able-to-drive-the-ball-275-yards-strong, should consider the gold tees at just over 7,000 yards, with a slope of 134. For most of the rest of us, the blue markers, 6,500 yards and sloped at 130, are plenty. Beginners and seniors would enjoy the white tees at just over 6,000 yards, but the forward markers will take the driver out of play much of the time for solid strikers.

The front nine offers a remarkable contrast between a typical golf course neighborhood and a dark foreboding wood. Within the space of a 30-second golf cart ride between green and tee, players will pass by the hammocks and swing set in somebody's backyard, and then immediately be on a wooden footbridge in semi-darkness, the overhanging trees blocking out the sunlight. Currently the ride between the sixth and seventh hole is a construction miasma, but there are plans in place for 50 townhouses to be built on the site, which features a large berm that abuts Interstate 16. The inward nine offers no such schizophrenia, and is a pleasant, consistent traverse through a standard golf course development.

The one-shot holes are straightforward but strong, playing between 130 and 190 yards from the blue tees. The par 4s are the strongest features on the course. Several are pronounced doglegs, requiring a tee shot that must bend to the contours of the fairway to maximize what little roll is available, and offer the best approach to the green. Said greens are generally large and quite undulating. A poorly struck approach that finds the wrong side of a swale will make 3-putting a distinct probability. The best hole on the outward nine is the 8th, a 380 yard

dogleg right, where a boldly struck tee ball must traverse a sizable water hazard that lurks on the right side of the fairway. The inward nine offers a mirror image on the 12th, where a well placed draw will allow a downhill, short iron approach on this par 4 of similar length.

"It isn't easy designing golf courses in the Coastal Empire and Lowcountry, because there's very little elevation change to work with," explains the architect. "You've got to create design features, and at Southbridge we created a diversified course with grass pockets, sculpted bunkers and berms. I think it's a good example of an inland, parkland-style golf course." Jones is correct in his assessment. His work in Savannah remains little-known mostly because it's overshadowed by all the great resort golf on Hilton Head and Sea Island more than any other explanation.

Jones remains self-effacing to the end, particularly when offering his final statement. "We attempt to incorporate the marshes and specimen trees into our work whenever we're able to. Creating courses in that area allows us to diversify our designs, and I love working there. I feel I've done some of my best work along the South Carolina and Georgia coast."

ESSAY: COUPLES GOLF CATACLYSM

Let's play a little multiple choice.
Which is the most excruciating ordeal?

A. Watching the entire Academy Awards telecast.
B. Waiting in the dentist's office for a root canal appointment.
C. Attempting to sit through the final foul-plagued minutes of an NCAA tournament basketball game.
D. Playing behind a mixed foursome on the golf course.

If you answered 'D,' congratulations. You win the grand prize, consisting of a year's supply of No-Doz, a Barcalounger and the magazine subscription of your choice.

I've got nothing against couples, I'm half of one myself. I certainly have nothing against golf, as I'm either playing, writing, or ruminating about the game virtually half my waking hours. What I have is a colossal problem with couples golf though, oftentimes a scenario that will stretch the benchmark time of four hours per round into some sort of lofty, unattainable goal like the three minute mile or four hundred yard drive.

Let me describe the tedium, this pathetic pas de deux I see stretching out in front of me: Ricky and Lucy ride together, followed by Fred and Ethel. While the girls recline in the passenger seat of their respective buggies, the boys are taking their rips from the men's tee box. After they hit, they return to their carts, put their drivers away and chauffer their sweethearts to the ladies tee some 20, 30 or 40 yards down the path. Now Lucy and Ethel leave the cart, grab their clubs, meander to the box, take a practice swing or three, then unleash titanic drives sometimes

more than 100 yards in length. They return to their shotgun position in the cart after carefully replacing their driver, and dutiful hubby resumes driving towards their ball.

Let me describe what I wish I could see: Instead of riding together, Edith and Archie split up, Archie riding with Meathead, Edith with Gloria. While Archie and Meathead are on the tee, Edith and Gloria are moving forward at an angle safely away from the intended line of flight, so the moment the second tee shot is either thundered or dribbled from the men's box, the girls are moving expediently towards the ladies tee box, driver in hand. If either Archie or Meathead is battling a serious case of the pull-hooks and caution is the watchword, then Edith and Gloria can wait laterally next to the men until the moment of impact, then proceed forward immediately.

Say it's a team competition, and Fred and Wilma insist on riding together, as do Barney and Betty. Then the ladies can walk forward at a prudent angle while the men hit, or take the carts forward themselves and let the men advance on foot to the red tees while their spouses are preparing to tee off. Either scenario works, and in addition to giving everyone the opportunity to walk a few paces and get a bit of exercise, should result in a time savings of at least a minute or two per hole. Just this simple arrangement alone will help turn a four-and-a-half hour marathon into a far more acceptable sub-four hour round.

As a general rule, I've seen far too much gallantry on the golf course regarding mixed foursomes. While I am personally acquainted with at least three couples where the wife is the superior player, it's still safe to assume that in the majority of cases the husband is the stronger/better/more experienced player. That said, there is no need to stand around interminably while a less talented spouse rainbows chip shots to and from the same bunker, or putts a 40-footer so hard it meanders into a lateral hazard near the green. Ready golf is an important element in mixed foursomes. Not only does it keep play moving, it lessens the pressure considerably on the player who, try as they might, continually remains the furthest from the hole.

I've got nothing against couples, I'm half of one myself.

It never ceases to amaze me that a couple who has lived together, supported each other, prospered together, raised children together, perhaps dance beautifully together, can be so out of step on the golf course. There's a simple recipe available that will insure a mixed foursomes match that can end well before sundown, and keep the patience and pulse rate of the groups following well within controllable limits. Use a minute amount of planning, a tiny bit of foresight and a dash of common sense. Stir together vigorously and let settle for a maximum of three hours and forty-five minutes.

PERSONALITY:
TIM O'NEAL – HE MISSED BY A WHISKER

The magnitude of the tee shot wasn't readily apparent to Savannah's Tim O'Neal. In December of 2000 he stood on the final hole of the final round on the final day of Q-School in Palm Springs, California. He had begun the day some four hours earlier, well back of the leaders. But O'Neal had played beautifully that afternoon, getting to eight under par on his round through fifteen holes. He knew he had vaulted back into contention for a PGA Tour card, and a bogey late in the round convinced him that a birdie on the last was a necessity to secure the coveted card.

"People talk about the pressure of Q-School," begins the soft spoken O'Neal, soon to be 30. "They talk about the six- round marathon and how grueling it is. But to me, getting though the second stage was the hard part. Once you're in the finals you'll be competing the next year, whether it's on the PGA Tour, Buy.com Tour, or with conditional status. I knew I'd have a place to play regardless, so I just relaxed, and went out and played."

It had taken the '97 Jackson State graduate four tries to make it to the finals. He won the Georgia State Amateur shortly after graduation, and then crushed the field by 17 shots at the prestigious Oglethorpe Invitational in Savannah. He turned pro, but missed Q-School semi-finals by a shot in that initial attempt. The next two years he made it through stage one but fell short at the semis, and finally broke through in 2000. In a field sprinkled with PGA Tour veterans the Savannah native played the last nine holes at six under par in the semi-final stage, and the resultant 29 exempted him into the finals with two shots to spare.

"I didn't do anything spectacular those first five rounds of the finals, just a shot or two under par every day," recalls O'Neal. "But I knew I could play that golf course." Unlike many of his competitors, the Johnson High School product was not examining the leader board obsessively. While many of his counterparts were making daily attempts to determine the scores they figured would be necessary to gain playing privileges, O'Neal was more laid back. "It was my first finals, and I wasn't studying the board closely at all. At the start of the last round I was well back in the field, and not even on the bubble, really. I just knew I needed to play well."

Play well he did, but that last salvo with the driver sailed far offline, over a bunker and into a lake. "It was a bad swing. I got the club stuck behind me, the face opened and I hit it into the water." O'Neal claims he never even considered the water a factor. In previous rounds he was right down the middle, and never gave the lake a second thought. "Maybe if I had known where I stood at the time I wouldn't have been so aggressive. I thought I needed birdie to get my card, when a bogey would've done it. There was no way to know though, because there are no scoreboards around to let you know where you stand."

He had to take a penalty drop near the hazard, some 250 yards from the green. His three wood sailed into a back bunker, leaving a downhill lie. The sand blast was long, a chip towards the flag and a missed putt from short range added up to a triple bogey seven. "I had no clue how close I was," says O'Neal, looking back with two years perspective. "The hole seemed to go by in slow motion, particularly after I hit it in the water. If I knew what I needed I might've laid up in front of the green with the third shot, and tried to get up and down with my wedge to secure my card." How did the catastrophe affect him? "The truth is, I didn't sleep that night," he admits. "That drive kept replaying in my head. All I could think about was that shot that went in the water."

It's a sad chapter but not a sad story, as O'Neal is blessed in many ways. He and his wife Melody live on Savannah's south side with their baby daughter Jordan. He's upbeat, in fantastic condition, has a ton of game and a supportive family. Beyond all that, he has a golf sponsor that's as generous as he is famous. O'Neal serendipitously hooked up with 'A' list celebrity Will Smith a few years back. The former "Fresh Prince" has been funding O'Neal's career, and allowing him to pursue his goal of eventually making it on Tour.

"I was in the right place at the right time," laughs the golfer. "I wasn't even planning on playing that week up in Richmond," he remembers, referring to a last minute decision to attempt to qualify for a Buy.com event.

It's logical to assume that Smith and O'Neal met in Savannah, when the Oscar nominee was in town filming *The Legend of Bagger Vance*, but it isn't the case. "I saw Robert Redford and Matt Damon out at Southbridge while they were filming in the area, but I never saw Will," explains O'Neal.

But Craig Crossley saw the Savannah product fire a smooth 66 up in Richmond. Crossley is a former PGA professional now serving as Director of Business Development for Will Smith Enterprises. O'Neal's eye-opening score was one thing, but not the only thing that impressed Crossley. "Tim's attitude and composure caught my attention," states the former Hampton University golfer. "He never hangs his head, or throws a club, or has anything but a smile on his face. Will has a fondness for helping people who have everything they need to succeed but the opportunity to prove themselves. Tim is a perfect example. He's a great player who hasn't had the chance to get to the next level."

"I got a call shortly after that Richmond event," remembers O'Neal. "Craig told me that Will Smith wanted to meet me. He flew me out to L.A. and I spent a week with him and his family."

O'Neal describes Smith's personality as similar to the TV character that made him famous. "He's really like the Fresh Prince of Bel-Air, especially on the golf course," states O'Neal. Although he learned the game at bedraggled Bacon Park in Savannah and comes from modest means, Tim O'Neal wasn't star-struck by his proximity to Smith. "I've been around some nice stuff and impressive people in the past," he claims. Not hard to understand, considering precocious golf ability opens doors in all directions that might otherwise be closed. "We hung out, got to know each other, and he decided to back me in my career."

Smith's sponsorship isn't based on 'return on investment.' Not only does he finance the airfare, rental cars, hotels, expenses and meals for O'Neal, but his staff makes all the arrangements as well. The golfer calls his liaison on Smith's staff, tells him where and when he needs to go to meet his instructor, or attempt qualifying or enter tournaments, and they call back with the itinerary completed shortly thereafter. It allows him to concentrate on his game without distraction. Smith also provides a swing coach, sports psychologist, strength and conditioning coach, practice facilities and a membership at Savannah Harbor. "Will Smith believes you should be the best you can be, regardless of occupation," says Crossley. "We're intent on providing all the tools Tim needs to become the champion he can be."

"There are no words I can use to describe what Will has meant to me," says O'Neal earnestly. "He's just a great guy. He's done this simply to see me succeed, and perhaps down the road I can repay him through earnings or endorsements, but I'm not obligated to. I have to say that if I were in his financial position I know I'd want to do the same thing for somebody else."

O'Neal made about $75,000 on the course in 2001, but missed retaining his Buy.com Tour card by less than $5,000. Claiming he golfed himself into exhaustion as the season wound down in attempting to maintain his playing privileges, he flamed out of Q-School in the second stage. The 2002 season was a patchwork of Canadian Tour events and attempts to Monday qualify for that week's Buy.com Tour event.

Once again, the second stage of Q-School proved to be his undoing. At the end of the 2002 season, he finished 13 shots out of a ticket to the final stage. But O'Neal remains undaunted, and looks forward to his next crack at the Qualifying Tournament.

"I'm committed to my career for as long as it takes. I know I'll make it on Tour, my time is coming," he concludes. "My mom said that things happen for a reason, and maybe I wasn't ready back in 2000. I do know I'll be back there again, and handle it differently. Next time I'm in the finals I will most definitely get through." His sponsor is steadfast as well, and Smith has told his protégé he'll stick with him "until the wheels fall off." Considering most every round he shoots is at par or better, the wheels are firmly affixed.

Essay:
Count 'em Up, Play 'em Down

Let me tell you a little about my brother-in-law John, and why he has a golf game worth admiring. He doesn't play often enough to have any type of established handicap, but since his career-best score for 18 holes hovers around 115, It's safe to assume he is somewhere between a 40 and a 45.

John swings the golf club like the proverbial rusty gate, carries no woods ("I have no idea where they're going to go ..."), and has a putter with a suction cup on the handle. He possesses neither power nor finesse, and exhibits negligible shot making or course management skills. As an added bonus, his etiquette is virtually nonexistent. So, you ask, what's the catch? John plays it as it lies, and counts every single shot.

I have remarked to him on more than one occasion that I've never seen such a combination of incompetence and accurate accounting. His response is simple and amazingly logical: "If I don't count every stroke every time, then how will I ever know if I'm getting any better?"

Pretty perceptive guy, this brother-in-law of mine. How many golfers do you know that beat the ball sideways, slapping it into every other hazard on the course, and then exclaim, "Put me down for a seven," or "Six is the most I can take."

Excuse me? The most you can take is the score that you shot on that hole. You adjust your score with Equitable Stroke Control when you enter it into the handicap computer after your round, not after you walk off the green when you just made a snowman.

I had the misfortune of being paired up with some over-zealous yahoo once, a nice enough guy, but a man who was rather pro-tective of his shiny new single-digit handicap. After watching him blow up on a hole or two on either side, he remarked to me how lately he had been on a hot streak, never scoring higher than 81 for almost a month. Of course, three or four times that afternoon he instructed me to "put down a 6—that's my limit." This delusional fellow had convinced himself that he was shooting near 80, when a thorough accounting would have revealed a score much closer to 90.

Why the hard line attitude? Because there is much more to golf than mist and mysticism, friendship, camaraderie, exercise, fresh air and all the rest of that happy horsemeat you used to see on Nike and Dockers commercials. Golf is also about numbers, plain and simple. Course ratings, par, handicaps, and ultimately how many blows it takes you to hole out 18 different times. Real golf is objective, not subjective. That's also why I admire John for always playing the ball down. This is in direct contrast to certain other in-laws in the family, speaking of in-laws, who have the tendency to get themselves out of more bad lies than Bill Clinton.

When you play the ball down, when you putt it out, when you eschew the mulligan, it makes the whole game more meaningful and clear-cut. The playing field is level; there is no nonsense or innuendo or gamesmanship. I play in a "blitz" once or twice a week, and it distresses me to often see the course marked for winter rules. To me, winter rules mean put away the clubs and go skiing. I don't like to roll my ball and I usually resist, though it puts my team at a competitive disadvantage. Everyone else is buffing it, shining it, and practically teeing it up on every shot. There's nothing like a solid game of "winter rules" to bring out a chorus of, "This is considered fairway, isn't it?" or, "Can I take a drop?" over and over as players look to gain any advantage, no matter how incremental, over the opposition.

I have remarked to him on more than one occasion that I've never seen such a combination of incompetence and accurate accounting.

I know golf is a difficult game to play well and most folks don't take it that seriously, they just want to have a good time. But counting them up and playing them down do not necessarily equate golf torture. The Scots who invented the game called it the "rub of the green;" we know it as the vagaries of the game. Sometimes your perfect tee shot ends up in a fairway divot. Other times your horrific slice that's heading out-of-bounds caroms off a tree and back to the middle. It's the nature of the beast. Take the good with the bad, and try to enjoy them both.

AREA GOLF COURSES

To paraphrase the late Mae West, "So many courses, so little time." The Carolina Lowcountry and coastal Georgia offer an abundance of golf that goes far beyond the subjective discussions in this book.

In the interest of providing a more thorough and comprehensive look at the entire scope of golf in the region, here is a listing of three dozen area courses that welcome guests.

Daufuskie Island Courses

Bloody Point	843-842-2000
Melrose	843-842-2000

Palmetto Dunes Resort

Arthur Hills at Palmetto Dunes	843-785-1138
George Fazio Course	843-785-1138
Robert Trent Jones Course	843-785-1138

Palmetto Hall

Arthur Hills at Palmetto Hall	843-689-4100
Robert Cupp Course	843-689-4100

Sea Pines Resort

Harbour Town Golf Links	843-842-8484
Ocean Course	843-842-8484
Sea Marsh	843-842-8484

Indigo Run

Golden Bear Golf Club	843-689-2200

Hilton Head Plantation

Country Club of Hilton Head	843-681-4653
Oyster Reef Golf Club	843-681-1745

Port Royal Plantation

Barony Course	843-689-4653
Planter's Row	843-689-4653
Robber's Row	843-689-4653

Shipyard Plantation

Shipyard Golf Club	843-689-4653

Bluffton

Crescent Pointe Golf Club	843-341-2500
Executive Golf Club	843-837-6400
Eagle's Pointe Golf Club	843-686-4457
Hidden Cypress Golf Club	843-705-4999
Hilton Head National Golf Club	843-842-5900
Island West Golf Club	843-689-6660
Okatie Creek Golf Club	843-705-4653
Old Carolina Golf Club	843-785-6363
Old South Golf Links	843-785-5353
Pintail Creek Golf Club	843-784-2426
Rose Hill Plantation	843-842-3740

Beaufort Area

Pleasant Point Plantation	843-522-1605
South Carolina National	843-524-0300

Savannah Area

Bacon Park	912-354-2625
Crosswinds	912-966-1909
Henderson Golf Club	912-920-4653
Lost Plantation	912-826-2092
Savannah Harbor	912-201-2007
Southbridge	912-651-5455
Wilmington Island Club	912-897-1615

*For more of Joel Zuckerman's writing visit
www.vagabondgolfer.com.*